Cell Metabolism: Growth and Environment

Volume I

Editor

T. A. V. Subramanian
Professor
Department of Biochemistry
Vallabhbhai Patel Chest Institute
University of Delhi
Delhi, India

CRC Press, Inc.
Boca Raton, Florida

Library of Congress Cataloging-in-Publication Data
Main entry under title:

Cell metabolism: growth and environment.

Bibliography: p.
Includes indexes.
1. Microbial metabolism. 2. Microbial metabolites.
3. Microbial synthesis. 4. Plants—Metabolism.
5. Metabolism—Regulation. I. Subramanian, T. A. V.
QR88.C45 1986 574.87′5 85-28966
ISBN 0-8493-5750-0 (v. 1)
ISBN 0-8493-5751-9 (v. 2)

Direct all inquiries to CRC Press, Inc., 2000 Corporate Blvd., N.W., Boca Raton, Florida, 33431.

International Standard Book Number 0-8493-5750-0 (v. 1)
International Standard Book Number 0-8493-5751-9 (v. 2)

Library of Congress Card Number 85-28966
Printed in the United States

PREFACE

The economic importance of several microbial metabolites, particularly antibiotics, has provided impetus to research efforts to elucidate the genesis of these compounds. The characterization and identification of biosynthetic routes has greatly facilitated the creation of newer and more efficient microbial products. The main difficulty is the tight degree of metabolic regulation which is frequently encountered in microorganisms, rendering them inefficient producers of metabolites. A certain degree of relaxation of the metabolic regulatory mechanisms is essential for economical production of an abundance of the desired microbial product. In other words, for the microbial cells to serve as efficient producers it is very important to circumvent or overcome the regulatory mechanisms involved. In the competitive world of commercial synthesis and production, simple molecules like glutamic acid and lysine are still made by fermentation rather than by chemical synthesis. About 300,000 tons of citric acid are produced annually from *Aspergillus niger* alone. In addition, microorganisms are extremely useful for carrying out bioconversion processes involved in the production of specific stereo compounds.

Increased biosynthesis of chlortetracyclin by developing mutants of *S. aureofaciens* resistant to the antibiotic has recently been reported. With the developments in biotechnology it is possible now to manufacture biologically important compounds like insulin on a large scale without depending on animal sources. Some of the hydroxylation reactions in steroid biosynthesis which require multistep procedures in chemical synthesis are greatly facilitated by microbial enzymes.

Historically, microbial products have been divided into primary and secondary metabolites. Primary products like proteins, nucleic acids, and carbohydrates are essential for growth and metabolism. Secondary products until recently were believed to have no significant role in the physiology of the producing organism and were considered as metabolic waste products.

As a first step it is necessary to find a wasteful strain which will overproduce and excrete a particular compound. The next stage is to modify the regulatory control of the strain to further increase its inefficiency. An example of such exploitation of microorganisms to waste their precious metabolites is seen in riboflavin production (up to 5000 $m\ell/\ell$) by commercial cultures of *E. ashby*.

These two volumes deal with the regulation of metabolism in microbes and plants with respect to differentiation, proteins, nucleic acids, aromatic acids, alkaloids, and other secondary metabolites like aflatoxins and streptomycin.

T.A.V. Subramanian
Vallabhbhai Patel Chest Institute
University of Delhi
Delhi-110007, India

THE EDITOR

Dr. T. A. V. Subramanian holds a Ph.D. in Biochemistry from the University of Madras. He has been Head of the Department of Biochemistry, V. P. Chest Institute, University of Delhi, since 1956.

Dr. Subramanian's training and experience, which includes 2 years at Columbia University, New York and 2 years at University of Wisconsin, Madison, is working in areas of biochemistry connected with metabolism of *Aspergillus parasiticus,* metabolism of various strains of tubercle bacilli, and pulmonary surfactants. He has more than 250 publications in these areas. He has been actively engaged in research on various aspects of the metabolism of *Aspergillus parasiticus* for more than 20 years. He has been President of the Society of Biological Chemists (India).

CONTRIBUTORS

B. K. Bhatnagar
Biochemistry Department
Vallabhbhai Patel Chest Institute
University of Delhi
Delhi, India

Zdeněk Hošťálek
Department of General Microbiology
Institute of Microbiology
Czechoslovak Academy of Science
Prague, Czechoslovakia

Martin Luckner
Department of Pharmaceutical Biology
Section of Pharmacy
Martin Luther University
Halle/Saale, East Germany

Jana Novotná
Department of General Microbiology
Institute of Microbiology
Czechoslovak Academy of Sciences
Prague, Czechoslovakia

T. Ramakrishnan
Professor
Microbiology and Cell Biology
Laboratory
Indian Institute of Science
Bangalore, India

Werner Roos
Department of Pharmaceutical Biology
Section of Pharmacy
Martin Luther University
Halle/Saale, East Germany

T. A. V. Subramanian
Department of Biochemistry
Vallabhbhai Patel Chest Institute
University of Delhi
Delhi, India

C. S. Vaidyanathan
Professor and Chairman
Department of Biochemistry
Indian Institute of Science
Bangalore, India

Eugene D. Weinberg
Professor of Microbiology
Department of Biology and
Program in Medical Sciences
Indiana University
Bloomington, Indiana

CELL METABOLISM: GROWTH AND ENVIRONMENT

Volume I

Protein Metabolism in Relation to Secondary Biosynthesis
Nucleic Acid Metabolism in Relation to Growth
The Spatial Organization of Secondary Metabolism in Microbial and Plant Cells
Aflatoxin Biosynthesis
Role of Oxygenases in the Metabolism of Phenolic Compounds
Regulation of Secondary Metabolism by Trace Metals
Index

Volume II

Biosynthesis of Polyketides
Secondary Metabolites of Fusarium Species
Genetics of Antibiotic Production
Formation Physiology of Ergot Alkaloids
Index

TABLE OF CONTENTS

Chapter 1

PROTEIN METABOLISM IN RELATION TO SECONDARY BIOSYNTHESIS

Jana Novotná and Zdeněk Hošťálek

TABLE OF CONTENTS

I. INTRODUCTION

Proteins are the most abundant molecules in the cell, making up 50% or more of the dry weight. They are found in all cellular components forming the basis of the cell structure and function. Each kind of protein is specialized for its biological function. They operate as enzymes, transporting and regulatory proteins, and also serve as structural and storage components. Many of them act in that part of metabolism which participates in the formation of secondary metabolites.

Complex, mutually related regulatory processes affect the amount of respective proteins as well as their activity at a number of levels. The scope and the variety of the regulatory mechanisms controlling the biosynthesis of secondary metabolites are believed to be similar to those controlling primary metabolism.[1-3]

The aim of this survey is to present the current knowledge of the relationships between protein metabolism and the biosynthesis of secondary metabolites. Apart from mentioning proteins as nutrients for producers, our attention will be directed to proteosynthesis ensuring the formation of various proteins, to the enzymes of secondary metabolism and the regulation of their synthesis, and, finally, to the alterations in protein metabolism associated with the resistance to autotoxic products.

Most of the examples have been selected from the kingdom of the streptomycetes, the important producers of the secondary metabolites. Nevertheless, the discussed concepts can often be extended to other producing microorganisms — bacteria and fungi.

II. PROTEINS AS NUTRIENTS

The complex media used in industrial fermentations also contain proteins (soybean flour, distiller's solubles). The use of such slowly utilizable sources of N and C is based on the experience of industrial microbiologists in obtaining considerably higher product yields when using these components in the medium than it is with ammonia, nitrate, and glucose (see Section IV). The streptomycetes are microorganisms capable of living on various rather complex substrates. For the utilization of proteins they produce extracellular proteases. Current biochemical knowledge of proteolytic complexes,[4-8] substrate specificities of the individual enzymes,[9-12] the regulation of their formation and secretion, and their roles in the sporulation[13] and selective protein degradation in streptomycetes[14] is not yet sufficient to determine the concrete regulatory relationships between this part of the producer metabolism and the biosynthesis of secondary metabolites.

III. PROTEIN SYNTHESIS

In this chapter we shall summarize data on the composition and the characteristics of the translation apparatus building a polypeptide chain from amino acids according to a mRNA template. To make the picture complete, attention will be paid also to mRNA synthesis in view of the importance of protein synthesis regulation at the transcription level. The relation of the own products to the proteosynthetic apparatus of secondary metabolite producers will be dealt with in Section V.

A. Incorporation of Amino Acids In Vivo

The incorporation of radioactive amino acids or labeled precursors of RNA synthesis was monitored throughout cultivation: *Streptomyces antibioticus* producing actinomycin,[15,16] *S. aureofaciens* producing tetracyclines,[17-20] *S. griseus* producing streptomycin[21] and candicidin,[22] and *S. hygroscopicus* producing turimycin.[23]

The rate of proteosynthesis reaches its maximum during the growth phase, then decreases steadily with the age of the culture. However, neither RNA synthesis nor proteosynthesis is discontinued in the course of intensive antibiotic production. A maximum in proteosynthesis in complex media was preceded by a peak in RNA production,[19,20,22,23] but in *S. aureofaciens* growing in semisynthetic medium or in *S. rimosus* and *S. granaticolor* growing in complex medium, no shift was observed between the maxima of RNA and protein synthesis.[19,24]

The synthesis of macromolecules at the transition period between spores and the vegetative form was determined in the synchronously germinating spores of *S. granaticolor* growing in a maximal synthetic medium.[25] RNA synthesis begins after 3 min of germination, proteosynthesis begins after 4 min, and DNA synthesis begins at the 60th to 70th min of germination. Rifampicin prevented ^{14}C-leucine from becoming incorporated into proteins. However, in *S. antibioticus*, Hardisson et al.[26] observed ^{3}H-leucine incorporation in minimal medium even after a rapid preparation of the spore suspension in the presence of rifampicin, which indicates the presence of a stable mRNA in dormant spores. The delay in the onset of the macromolecule synthesis in the germinating spores of *S. antibioticus* observed in complex medium has been explained by a dilution of the radioactive exogenous precursors with the cold precursors contained in the medium.[26,27]

Guijarro et al.[28] dealt with the stability of the RNAs synthetized during germination and determined the half-life of mRNA formed both during the germination process and vegetative growth of *S. antibioticus*. The obtained values (4 to 6 min) for the spores germinating for over 2 hr as well as for the vegetative mycelium were higher in comparison with those obtained for *Bacillus subtilis, B. thuringiensis,* and *Escherichia coli,* which may be the result of a rather slower growth rate in streptomycetes. Spores at the initial stage of germination exhibited the greatest RNA stability (half-life of 22.5 min).

B. Proteosynthetic Apparatus

The individual components of the proteosynthetic apparatus of streptomycetes were investigated and characterized in detail mainly in conjunction with their resistance to their own products (Section V) or when studying lifecycle regulation at the translation level.[29-31]

A comparative study of *S. granaticolor* ribosomes was performed in the vegetative cells and dormant spores.[30,32] The spores contain free ribosomes; the presence of polysomes is unlikely. Qualitative changes in the electrophoretic pattern of the ribosomal proteins in the spores and the vegetative cells were observed. The number of the spore ribosomal proteins is lower.[32] Purified 30S and 50S subunits from the mycelium contain 21 and 36 ribosomal proteins, respectively, in *S. granaticolor*.[32] Only eight proteins have their electrophoretic mobilities identical with the corresponding ribosomal proteins of *E. coli*.

S. aureofaciens has 21 and 35 proteins in the 30S and 50S subunits, respectively. Separated ribosomal proteins from the 50S subunits of the ribosomes were used for the identification of structural homologies with the 50S proteins of *E. coli* by means of immunological techniques. About one half of these proteins shared common immunochemical determinants with the corresponding proteins of the 50S subunits of *E. coli*.[33] The SL5 protein from the large ribosomal subunit of *S. aureofaciens* is also functionally analogous to the L7/L12 protein of *E. coli* as revealed in the reconstitution studies of the deficient ribosome of *E. coli*.[33]

Ribosomes from the mycelium of *S. aureofaciens* producing tetracycline differ in the electrophoretic mobilities of the ribosomal proteins S2, S10, and L19 from the corresponding proteins of the same strain where the tetracycline production was suppressed by a modification of the cultivation conditions.[34] The changes observed in the positions of the spots are probably the result of post-translation modifications of the proteins.

Ribosomes are subject to some structural modifications in the course of differentiation of *S. griseus,* which is demonstrated by different electrophoretic profiles of the ribosomal proteins from spores and vegetative mycelium. The proteolytic activities of the suspensions of washed ribosomes from vegetative hyphae are higher than those of the ribosomes from the spores.[35]

The differences in the ribosome sedimentation characteristics and in the content of the ribosomal proteins of *S. granaticolor* spores and the vegetative cells indicate that the majority of the ribonucleoprotein particles present in dormant spores are ribosomal precursors.[36]

Ribosomal RNAs of *S. griseus* were isolated and characterized for their sedimentation coefficients (16S and 23S),[37,38] thermal stability, and nucleotide composition.[38] Electrophoretic mobilities of the 23S and 16S rRNAs from the vegetative cells and the spores of *S. griseus* were identical, but the 23S rRNA of the streptomycete ribosomes migrated with a much lower speed on polyacrylamide gel than did the 23S rRNA of *E. coli* ribosomes.[39]

In terms of antibiotic production, it is of significance that the streptomycete rRNA can be modified by the methylation of adenine or ribose, which was discovered in connection with a resistance to some antibiotics with their target sites on ribosomes[40-45] (see Section V).

While studying other components of the streptomycete translation apparatus it was found that the initiation factor IF3 from *S. aureofaciens* can be substituted for the IF3 of *E. coli* in in vitro initiation with Qβ RNA as the template and with the ribosomes of *E. coli*.[19] The ribosomes of *S. aureofaciens* were less active in the formation of the initiation complex than those of *E. coli*. This is attributed to a probable binding of tetracycline. By comparing the dissociation of the 70S ribosomes by means of sucrose density gradient, it was shown that the ribosomes of *S. aureofaciens* required more IF3 for the dissociation than did the 70S ribosomes of *E. coli*.

The elongation factor EF-Tu was purified from the mycelium of *S. aureofaciens* and compared with the EF-Tu of *E. coli*.[46,47] Regardless of the differences in the molecular weights and their thermal stability, the two factors are structurally related as follows from their identical electrophoretic mobilities on SDS gels and their immunological similarity. They are even functionally interchangeable with the ribosomes of *E. coli* in the poly (U)-directed polymerization of phenylalanine tRNA.[19]

DNA-dependent RNA polymerase was purified about 700 times from the 12-hr cells of *S. antibioticus*, and 400 times from the 48-hr cells.[48] Both enzymes contain subunits corresponding to the β-, β′-, and α-subunits of *E. coli* RNA polymerase, but on polyacrylamide gel no band corresponding to the α-subunit was observed. In addition, in the 12-hr RNA polymerase there is another protein present (45,000 daltons) in a relative molar concentration to the subunit 0.4. The polymerases from the 12- and 48-hr cells somewhat differ in their template specificities: the 48-hr polymerase exhibits a higher activity with calf thymus DNA in comparison with other natural DNAs, and it is more active with poly-(dA-dT) than the polymerase from young cells.[48] Partially purified RNA polymerases from *S. griseus* and *S. mediterranei* exhibited rather weak transcription activities when compared with an analogous preparation from *E. coli*.[49]

C. Protein Synthesis In Vitro

In vitro translation systems are important means for studying the regulation of protein synthesis whether the lifecycle or the resistance of the producers to their own toxic products is being investigated. An evaluation of the efficiency of such systems always demands certain caution regarding the effects of endogenous proteolytic activity.[30,50] The age of the culture from which the cell-free extracts or the individual components of the systems were prepared was shown to be the decisive factor for their activities.[29,31] These results indicate that in the course of the lifecycle, ribosomes are subject to such structural changes which apparently lead to a reduction in their affinity to mRNA.

The activity of translation systems in vitro derived exclusively from streptomycetes is usually considerably low when compared with other bacteria.[31,34,51-53] Therefore, it is usual to combine various hybrid systems composed of the components isolated from the streptomycete under study and from other organisms. At the same time, advantage is taken of a higher efficiency and eventually it is possible to avoid some characteristics of the components potentially interfering with the problem under investigation.

To study the mechanism of fructokinase induction by fructose, an in vitro system was employed to synthesize fructokinase with the S-30 fraction from *E. coli* and RNA prepared from *S. violaceoruber*.[54] The synthesized fructokinase was identified by its catalytic activity at pH 10 and by analogous specificity of the native enzyme and of the enzyme synthesized in vitro. Its presence among the radioactive products of the protein synthesis was confirmed by the transition of the molecular weight from 20,000 (monomer) to 80,000 (tetramer) in the presence of Mg-ATP and by electrophoresis of the purified preparations. The presumptive mRNA for the fructokinase apparently contained poly(A) sequences of variable length as assumed by the authors from the poly(U)-Sepharose binding assay of RNA. In the course of the cultivation of *S. violaceoruber*, a suitable moment could have been selected for the extraction of RNA to ensure about 7% of the total incorporated amino acids for an individual protein — the fructokinase.

As another example of the utilization of polyribosomes from the cells of *S. antibioticus* of a suitable age (containing mRNA for an enzyme of the secondary metabolism), we shall mention the cell-free synthesis of a subunit of phenoxazinone synthase, the enzyme catalyzing the formation of actinomycin chromophore.[55] The author isolated polyribosomes from an 18-hr mycelium. An analysis employing a sucrose gradient revealed that the polysome preparation consisted mainly of mono- to tetrasomes, although larger polysomal material was also present. The polysomes of *S. antibioticus* were active in cell-free proteosynthesis. Amino acid incorporation in the cell-free system was stimulated by the addition of *E. coli* tRNA, spermidine, and factors washed off the ribosomes of *E. coli*. Under optimal conditions, the polysomes of *S. antibioticus* synthesized a number of proteins, one of which had the molecular weight of 100,000 like that of the phenoxazinone synthase subunit. The synthesis of this protein was verified by an immunological analysis of the products using an antibody against purified enzyme. Electrophoretic analysis of the immunoprecipitates demonstrated the presence of one radioactive protein migrating approximately in the same way as did the marker of the authentic phenoxazinone synthase. In spite of discrepancies in the accurate molecular weights of the product of the cell-free proteosynthesis and the subunit of the authentic enzyme, it was demonstrated that the protein exhibiting immunological and electrophoretic characteristics of phenoxazinone synthase, though enzymatically inactive, was synthesized in vitro with polyribosomes from *S. antibioticus*. It was noted in these experiments that about 20% of the total radioactivity incorporated by the 18-hr polysomes after 40 min of the incubation was specifically precipitated by the antiserum.

In vitro protein synthesis directed by mRNA and DNA was also performed in *S. aureofaciens*.[56] The translation products were traced by autoradiographic detection after a separation on polyacrylamide gel and have not yet been characterized. A considerable reduction in the number of proteins synthesized in the presence of chlortetracycline was observed.

IV. ENZYME SYNTHESIS AND SECONDARY METABOLISM

A. Enzymes of the Secondary Metabolism

Intensive research carried out during the past decade brought the first, though fragmentary and incomplete, data about enzymes or enzyme systems participating in the biosynthesis of secondary metabolites in streptomycetes.[57] The enzymes of transformation of the streptidine moiety of dihydrostreptomycin in *Streptomyces humidus* or of the bluensidine moiety of bluensomycin in *S. globosus* were studied in detail.[58]

The diversity of the structures as well as of the biosynthetic pathways of the secondary metabolite formation make it impossible to create a comprehensive system to classify the enzymes of the secondary metabolism; they can be divided into two principal groups. First, there are the intrinsic enzymes of the secondary metabolism which modify the structure of the so-called primer or the parent biosynthetic intermediate. This compound is usually formed by the condensation of the primary precursors which is catalyzed by a specific enzyme complex. The other group includes enzymes participating in the formation of the primary precursors of the secondary metabolites. Apart from these systems, tens of associated anaplerotic enzymes can be considered which provide energy, reduction or oxidation equivalents, or the donors of the particular functional groups substituting the parent intermediate.[59]

1. Enzymes Catalyzing the Formation and Modification of the Parent Intermediate

The enzyme systems of the secondary metabolism have been studied mainly in compounds of known structure and biogenesis. Oligoketides, the substances formed by the condensation of acyl-CoA building units, are a relatively well-known group of secondary metabolites. A classical example of the oligoketide biosynthetic pathway is the transformation of 6-methylsalicylic acid, the parent intermediate of patulin in *Penicillium urticae.*

The first step of the biosynthetic pathway is the condensation of one acetyl-CoA molecule and three molecules of malonyl-CoA to give 6-methylsalicylic acid,[60-62] which is then transformed to patulin and some by-products via a series of further enzyme reactions.[63-70]

The course of the biosynthetic pathway of the tetracycline antibiotics is similar, though much more complex. The tetracycline skeleton condensation is assumed to take place on a hitherto hypothetical multienzyme complex called anthracene synthase. Vaněk et al.[59] presuppose the sequence of 11 reactions resulting in the transformation of tricyclic nonaketide, formed by a condensation, into the final product, chlortetracycline.

Three enzymes catalyzing the biosynthetic steps of the tetracycline antibiotic series have been identified so far. The double N-methylation at C_4 is catalyzed by *S*-adenosylmethionine: dedimethylamino-4-aminoanhydrotetracycline *N*-methyltransferase.[71] The enzyme was detected in cell-free extracts from various strains of *S. aureofaciens* and *S. rimosus.*

Běhal et al.[72] described anhydrotetracycline oxygenase transforming anhydrotetracycline to dehydrotetracycline in *S. aureofaciens.* The hydratation of anhydrotetracycline requires NADPH and O_2. Eighty percent inhibition of the purified enzyme was achieved by 1 m*M* chlortetracycline, while its activity could be stimulated by 10^{-4} to 2.10^{-3} *M* Ca^{2+} and 10^{-4} to 10^{-2} *M* Mg^{2+}.[73] Finally, NADP: tetracycline 5a(11a) dehydrogenase, the final enzyme of the reaction sequence, was described[74] carrying out the reduction of 5a(11a) double bond.

A number of the secondary metabolite molecules are composed of moieties which have been the products of various biosynthetic pathways. The final stèps of the biosynthesis are then catalyzed by specific synthetases (ligases) joining together the individual building blocks. Such an enzyme is, e.g., novobiocic acid synthetase in *S. niveus,* which combines via the amide bond (in the presence of ATP) the A ring (substituted benzoic acid) with the ring B (the coumarin moiety) of the novobiocin molecule.[75]

2. Enzymes Catalyzing the Formation of Primary Precursors

Metabolic pathways yielding identical biosynthetic intermediates for both primary and secondary biosynthesis are generally supposed to be identical. These conceptions were confirmed experimentally while studying malonate and methylmalonate formation. In *S. noursei* var. *polifungini,* the rates of nystatin A_1 biosynthesis and lipogenesis are controlled by the level of carboxylated acyl-CoA units in the common pool.[76] In *S. hygroscopicus,* too, malonyl-CoA and methylmalonyl-CoA are generated via the classical carboxylation system of acetyl-CoA and propionyl-CoA, respectively, and the rate of turimycin formation is directly proportional to the intensity of these reactions.[77]

This conception is contradicted by the results of Běhal and Vaněk,[78] who found in *S. aureofaciens* a very low activity of acetyl-CoA carboxylase during chlortetracycline biosynthesis in the production phase in contrast to that found in the vegetative growth phase. Běhal et al.[79] investigated the activities of the enzyme systems participating in the two possible alternative pathways of malonyl-CoA formation from the common precursor, phosphoenolpyruvate. On the one hand, there is pyruvate kinase and the pyruvate dehydrogenase complex providing acetyl-CoA (which is later carboxylated); on the other hand, there is phosphoenolpyruvate carboxylase and the enzyme system decarboxylating oxalacetate to malonate.[80]

The activities of pyruvate kinase are really the highest in the exponential phase. Conspicuously low activities of these enzymes were then found in the production variant at the time of the antibiotic formation. On the contrary, phosphoenolpyruvate carboxylase activity in this strain was increasing in the stationary phase, and this enhancement was in correlation with an increased rate of chlordetetracycline synthesis. A decarboxylation of 4-^{14}C-oxalacetate to malonate was demonstrated in a cell-free extract using gas chromatography. The reaction required the presence of CoA and NAD^+.

Experiments involving the incorporation of 2-^{14}C-acetate, 3-^{14}C-pyruvate, and 2-^{14}C-malonate into the antibiotic molecule also demonstrated the existence of the alternative pathway of malonyl-CoA formation. The radioactivity of the carbamoyl group (as determined by Hofmann degradation) following the incorporation of the labeled substrates was roughly identical — 2.8%. This indicates a high degree of randomization since the carbamoyl group is, in origin, the caboxyl group of malonate (acetate) and therefore it should not have been radioactive at all, provided that malonate was formed from acetate by direct carboxylation.

The results mentioned above brought an essential finding that malonyl-CoA used for the biosynthesis of a secondary metabolite in the stationary phase can be formed by a pathway alternative to that supplying the identical intermediate of fatty acid biosynthesis in the exponential phase. This finding was confirmed by Venkitasubramanian et al.[81] in their study of malonyl-CoA formation in the aflatoxin producer *Aspergillus parasiticus.*

3. Regulation of the Synthesis of Enzymes of the Secondary Metabolism

Under nonlimiting conditions, the secondary metabolic pathways in standard strains make up usually just a small fraction of the metabolic activity. The regulatory mechanisms prevent the metabolite overproduction and secure efficient nutrient utilization. Only extreme cultivation conditions or hereditary changes favoring the activity of the secondary metabolic pathways make a major part of the metabolic activity proceed in this direction. The spectrum of the regulatory mechanisms of the biosynthesis of the secondary metabolism enzymes is broad and diverse. Usually it is the question of such regulatory mechanisms like induction, catabolite regulation, or a repression by the end product.

A classical example of induction, i.e., control of enzyme synthesis by the substrate or its analog, was described by Krupinski et al.[82] in *Claviceps.* Tryptophan (as well as thiotryptophan and 5-methyltryptophan) induced *de novo* synthesis of dimethylallyltryptophan synthase, the initial enzyme of the alkaloid biosynthetic pathway. A similar effect was noted for methionine (or norleucine) in cephalosporin C biosynthesis in *Cephalosporium acremonium*[83] or during the phosphomycin biosynthesis in *Streptomyces fradiae.*[84]

An important part in the induction of the enzymes of the secondary metabolism is also apparently played by endogenous inducers such as the A factor in the synthesis of streptomycin[85] or the endogenous effector of staphylomycin biosynthesis in *S. virginiae.*[86] The A factor induces the synthesis of the streptomycin biosynthesis enzyme apparatus[87] and at the same time affects the development and morphological differentiation (without changing the genotype) in *S. griseus.*[88] A particularly conspicuous manifestation of its effect is in the nonproduction asporogenic mutants where it accounts for a phenotypic reversion — the formation of aerial mycelium and spores.[89] This factor was identified as 2S-isocapryloyl-3S-hydroxymethyl-γ-butyrolactone.[90,91] Its presence was confirmed in many actinomycete species.[92] The presence of anthracycline formation inducers was demonstrated in strains of *S. griseus.*[93] While studying its chemical nature, Gräfe et al.[94,95] confirmed the primary importance of the butyrolactone moiety with its aliphatic side chain containing a polar substituent for the biological activity of the so-called endogenous inducers of secondary metabolism.

The inhibitory effect of the rapidly metabolized carbon sources upon an antibiotic biosynthesis has been usually characterized as a catabolite regulation of the secondary metabolism enzymes.[3,96,97] As long as the antibiotic production alone was monitored, disregarding the specific activities of the relevant synthetases, the character of the inhibitory effect is difficult to classify. Thus, Demain[96] described an inhibitory effect of glucose on penicillin production by the resting cells of *Penicillium chrysogenum* in which the system of the antibiotic biosynthesis was fully developed. This inhibitory effect of glucose was similarly noted in the case of neomycin[98] or siomycin[99] synthesis. Long-term incubation of the mycelium and the possibility of *de novo* protein synthesis do not enable one to decide whether it was a catabolite inhibition or the repression of the synthesis of the antibiotic pathway enzymes.

Similar results were obtained in *S. clavuligerus* in the study on the catabolite regulation of cephalosporin production.[100] Here, the type of catabolite regulation is not clear either, but this effect was shown not to be the result of such trivial stimuli as an acid production or oxygen limitation, but that a real regulation effect was dealt with.

Another experimental model used for studying the catabolite regulation was the biosynthesis of cephalosporin C and penicillin N in *C. acremonium.*[101] The specific production of β-lactam metabolites reached its peak on the slowly assimilated sugars such as lactose or sucrose. The antibiotic production took place also during the growth phase on glucose when used in a low concentration. After the depletion of glucose from the medium, the production rate sharply increased. At a high initial glucose concentration no β-lactams were produced.

The inhibition effect of glucose added to the culture was described for *S. antibioticus.*[102] Actinomycin production was inhibited and the synthesis of phenoxazinone synthase (the enzyme catalyzing the synthesis of the actinomycin chromophore) was repressed. The terminal enzyme of the biosynthetic pathway of puromycin, *O*-demethylpuromycin-*O*-methyltransferase, is repressed by glucose present in the medium at the beginning of the cultivation as well as by its addition to the growing culture.[103]

The biosynthesis of a number of secondary metabolites is inhibited by mechanisms resembling end-product inhibition.[104] Chloramphenicol represses the formation of arylamine synthetase, an enzyme catalyzing chorismic acid amination to an aromatic amine, which is an intermediate product of the chloramphenicol biosynthesis.[105] End-product repression is likely to block enzyme synthesis at the time of maximum velocity of the secondary biosynthesis. This results in a gradually decreasing concentration of the appropriate synthetase and a slow down in the formation of the product. This phenomenon was observed for the gramicidin synthetases in *Bacillus brevis,*[106] novobiocic acid synthetase in *S. niveus,*[107] or dimethylallyltryptophan synthase in *Claviceps.*[108]

4. Regulatory Role of Phosphate Ions

The formation of a number of secondary metabolites is inhibited by the presence of phosphate ions in the medium even at their suboptimal concentrations for the growth of the production microorganism. The inhibitory effect of P_i in aminoglycoside biosynthesis can be accounted for by the inhibition of both the synthesis and the activities of the phosphatases participating in the transformation of the streptidine moiety of dihydrostreptomycin.[58] Two specific repressible guanidinocyclitol phosphate phosphatases have been described.[109] Also, in *S. fradiae,* the synthesis of the specific phosphatases participating in the biosynthesis of neomycin is repressed by both P_i and glucose.[98]

In those cases where the biosynthesis of a substance occurs without the participation of phosphorylated intermediates, the synthesis of the enzymes of the secondary biosynthesis usually becomes derepressed as a result of phosphate depletion from the medium. While studying the effect of orthophosphate on candicidin biosynthesis in *S. griseus,* Martín et al.[110] came to the idea that P_i inhibits the activity of candicidin synthetase and probably represses its synthesis at the transcription level. Inorganic phosphate represses the synthesis of anhydrotetracycline oxygenase in *S. aureofaciens,*[111] and this inhibition effect could be abolished adding benzyl thiocyanate into the fermentation medium. Madry and Pape[112] observed inhibition of the synthesis of *S*-adenosyl-L-methionine:macrocin *O*-methyltransferase by phosphate ions. The enzyme acts at the last step of tylosine biosynthesis.[113,114] Phosphate also inhibited the synthesis of two other enzymes that take part in the formation of the mycarosyl moiety of tylosin.[112]

Possible mechanisms of the phosphate ion inhibition effect on the biosynthesis of secondary metabolites have been discussed by Martín in his extensive survey.[115] There is a large body of experimental evidence that P_i favors the metabolic pathways of the primary metabolism, namely the catabolism of carbohydrates, thus limiting the synthesis of the inducers or precursors of the secondary biosynthesis. Therefore, an equilibrium between the degree of expression of the growth-related processes and the processes of the secondary biosynthesis is obviously controlled by P_i concentrations. However, the way of signal transmission and the nature of the effector substance controlling the expression of the secondary metabolism are quite obscure.

Hošťálek et al.[116] found that a decrease in the level of adenylates in the culture of *S. aureofaciens* is accompanied by the formation of anhydrotetracycline oxygenase. Intensive biosynthesis of chlortetracycline takes place during this period until the stage is reached when a repeated increase in the adenylate concentration becomes accompanied by a decrease of the enzyme level and the antibiotic biosynthesis is brought to a standstill. At a higher phosphate ion concentration (+1 μmol P per milliliter), there is no reduction of the level of adenylates,[117] and the synthesis of anhydrotetracycline oxygenase becomes repressed.[111]

Vu-trong et al.[118] also reported that the expression of tylosin formation in *S. fradiae* occurs following a decrease in the level of total adenylates; the culture begins to synthesize the enzymes of propionyl-CoA conversion (propionyl-CoA carboxylase) and the conversion of methylmalonyl-CoA (methylmalonyl-CoA carboxytransferase). Glucose addition to the culture increased adenylate levels, suppressed the synthesis of both enzymes, and reduced the yield of the antibiotic.

The mechanism of the regulatory effect of the level of adenylates is not understood at present. However, the above findings point to a common character of the two hitherto independently considered phenomena: catabolite regulation of the secondary metabolism by rapidly assimilable carbon sources, as well as its inhibition by increasing concentration of phosphate ions.[119] In both cases, the level of total adenylates plays a significant part in the expression of the secondary metabolism.

B. Other Enzymes Related to Secondary Product Formation

A significant factor affecting the rate of the formation of the secondary metabolites is the availability of the building units, including functional groups, redox equivalents, etc., from which the individual metabolites are synthesized. These are usually common for both primary and secondary metabolism. Therefore, regulation of the synthesis and activity of enzymes catalyzing a large variety of primary metabolism reactions is an inseparable part of the considerations on protein metabolism and the biosynthesis of secondary metabolites.[59,120]

The tricarboxylic acid cycle is a metabolic system competing with tetracycline biosynthesis.[121] A characteristic feature of the high-production strains of *S. aureofaciens* is the low activity of the enzymes of the cycle in the phase of intensive antibiotic biosynthesis which makes it possible to utilize the C_3 components in chlortetracycline formation.

Another example of decreasing the activity of enzymes belonging to the pathways competing with those of antibiotic biosyntheses could be the feedback regulation of the first enzymes of each branch leading from chorismate to specific aromatic amino acids, tryptophan and phenylalanine, in *S. venezuelae*.[122,123] Chorismate thus may be efficiently used for the biosynthesis of chloramphenicol.

The initial part of a biosynthetic pathway may be common for the production of both primary and secondary metabolites. In some cases a primary end product decreases by the mechanism of feedback regulation the synthesis or the activity of enzymes that operate in the common part of the branched pathway. Nevertheless, such a regulation by aromatic amino acids need not always be present as shown by the observations of 3-deoxy-D-*arabino*-heptulosonate-7-phosphate synthetase (DAHP) of *S. venezuelae*, an enzyme catalyzing the earliest step of the shikimate pathway.[124]

Mutations making the enzyme(s) of the common part of the pathway released from such a feedback regulation may lead to an overproduction of the antibiotic. *S. clavuligerus*, a cephamycin C producer, was successfully manipulated in this way to possess a mutated aspartokinase that was no more inhibited by the concerted effect of lysine and threonine.[125]

There is evidence for coregulation of the nitrogen metabolism and of the antibiotic synthesis which may be decreased by the addition of NH_4^+ or nitrate.[107,126-136] Glutamine synthetase, a central enzyme of nitrogen metabolism, was studied in *S. noursei*,[130] *S. clavuligerus*,[137] *S. cattleya*,[138] and *S. venezuelae*[133] with respect to the synthesis of nourseothricin, cephalosporin, thienamycin, and chloramphenicol, respectively.

Streicher and Tyler[139] found a 20-fold difference in the glutamine synthetase protein levels between NH_4^+ repressed and derepressed cultures of *S. cattleya*. The data do not make it possible to decide whether the regulation occurs at the level of transcription of the *glnA* gene, or at the level of velocity changes in the degradation of either the appropriate mRNA or of the enzyme subunits.

However, the activity of an enzyme can be regulated by modifying the activity of the synthesized enzyme as well. The addition of NH_4^+ resulted in an immediate inactivation of glutamine synthetase of *S. cattleya*.[138,139] Streicher and Tyler[139] demonstrated that the enzyme activity is regulated by adenylation in a way analogous to that in enteric bacteria. This covalent modification was similarly demonstrated in vitro after the addition of ATP to the crude extract. Several other proteins in *S. cattleya* were also found to incorporate ^{32}P, and a phosphodiester bond was demonstrated in some of these by the action of snake venom phosphodiesterase.[139] This would suggest an analogous mode of regulation of the enzyme activities foreseen for these hitherto unidentified proteins.

A high glutamine synthetase level was found to be connected with the antibiotic production,[133,137,138] However, the lack of a consistent correlation between the glutamine synthetase level and the onset of the antibiotic synthesis in *S. venezuelae* does not make this enzyme a suitable candidate for the probable specific effector of the secondary metabolite processes.[133] Besides, a modern view of glutamine synthetase in enteric bacteria no longer claims its central regulatory role in nitrogen regulation,[140] which also supports the idea of an incidental occurrence of the high glutamine synthetase activity and of the overproduction of antibiotics.

Our purpose was to outline the broad scale of actions of primary metabolism on the biosynthesis of secondary metabolites. Though many essential facts concerning the mutual relations of the primary and secondary metabolism have emerged during a few recent years, most of the high-producing strains still remain to be characterized in detail from the viewpoint of the target metabolic steps that had been affected in eliciting overproduction.

V. RESISTANCE OF THE PROTEOSYNTHETIC APPARATUS TO AUTOTOXIC SECONDARY PRODUCTS

Significant changes in the protein metabolism of producing organisms occur in association with the development of resistance to their own products. Many antibiotics exert their effects directly on the proteosynthetic apparatus even in their own producers. Resistant microorganisms, in general, may employ several possible mechanisms to prevent harmful drug effects.[104,141,142]

A modification of the target site in such a way that the antibiotic molecules cannot exert their inhibitory effects is one of the modes of self-defense. Another possibility is a detachment of the target site to prevent it from getting into contact with the antibiotic either in space or in time (compartmentation, exclusion or expulsion of antibiotics out of the cells, antibiotic synthesis only after the termination of growth). Finally, the third possibility is an inactivation of the antibiotic molecule by the action of enzymes modifying the substance to become a harmless derivative.

This review will be aimed primarily at those alterations of protein metabolism that are associated with the resistance to its own products affecting the proteosynthetic apparatus.

A. Modification of the Target Site

1. Peptide Antibiotics

The resistance to thiostrepton in the producing organism *Streptomyces azureus* is a very well-documented example of a target site modification.[40,41] The antibiotic inhibits protein synthesis in the sensitive bacteria by an irreversible binding to the 50S ribosomal subunit; it acts upon the processes associated with the GTP hydrolysis center of the ribosome, including the binding of guanine nucleotides, elongation factors, and tRNA.[143] Thompson et al.[144] demonstrated that the binding site for thiostrepton can be constructed in vitro when using only the 23S ribosomal RNA and the ribosomal protein L11.

In *S. azureus,* there is the enzyme methylase that specifically methylates the 23S ribosomal RNA producing a single residue of 2′-*O*-methyladenosine, thereby rendering the ribosomes unable to bind thiostrepton. This methylase of thiostrepton resistance was purified 11,000 times and its substrate specificity, approximate molecular weight, and optimal conditions for its activity were determined.[42] The enzyme uses *S*-adenosylmethionine as a cofactor; its preferred substrate is a phenol-extracted 23S RNA. Complete 50S ribosomal subunits are not methylated at all. The protein L11 suffices to block the effect of the enzyme on the 23S RNA, which indicates that the site of methylase action and the binding site for L11 could overlap.[42]

Based on the fact that naked 23S rRNA (but not the 50S ribosomal subunit) is the substrate for methylase, Thompson and Cundliffe[42] presume that the enzyme acts on RNA in an early stage of its existence, possibly at the level of a primary transcription product. The ribosomes of *S. azureus* are resistant to thiostrepton throughout the whole lifecycle and not just in the phase of antibiotic production, which indicates a constitutive methylase formation.

Also, in other antibiotic producers, *S. laurentii* (thiostrepton), *S. sioyaensis, Planomonospora parontospora,* and *S. actuosus* (siomycin, sporangiomycin, and nosiheptid, respectively; antibiotics structurally more or less related to thiostrepton), it is very likely that RNA pentose methylation giving rise to one 2′-*O*-methyladenosine in the 23S rRNA is involved in the resistance to their own products.[52,145]

Even the berninamycin producer *S. bernensis* has a similar ribosomal RNA methylase that is able to modify ribosomes to make them resistant to berninamycin[146] as well as to thiostrepton and micrococcin, although berninamycin, a cyclic peptide antibiotic, is structurally different from thiostrepton. However, Thompson et al.[146] showed that berninamycin binds to the 23S RNA complex with protein L11 in a way analogous to thiostrepton, even though the relationship between their binding sites has not yet been exactly determined, and berninamycin also affects the function of the ribosomal site A. The rRNA methylase of *S. bernensis* exhibits analogous substrate specificity and produces identical methylated product (with 2′-*O*-methyladenosine) as the enzyme of *S. azureus.*[146]

2. Macrolides-Lincosamid-Streptogramin Type B (MLS)

Just a little different mechanism of the modification of the 23S rRNA target site occurs in *S. erythreus,* which produces the macrolide antibiotic erythromycin.[43,45] When comparing the ribosomes of *S. erythreus* and *S. griseus,* Teraoka and Tanaka[147] found the ribosomes of *S. erythreus* lacking an affinity to erythromycin and also resistant to other macrolides (leukomycin, spiramycin, tylosin) and to lincomycin. 23S RNA of *S. erythreus* contains N^6,N^6-dimethyladenine.[43] This fact, along with the previous findings on the inducible methylation coded for by the plasmid in staphylococci, has led to the idea about the resistance mechanism in *S. erythreus* being similar to that in staphylococci.

Skinner and Cundliffe[45] proved this hypothesis: they isolated and characterized in vitro the enzyme erythromycin resistance methylase, responsible for the presence of dimethyladenine in *S. erythreus* 23S rRNA, which rendered the ribosomes and their production organism resistant to erythromycin and to other macrolides, lincosamid, and streptogramin type B (MLS resistance). It certainly will be interesting to find out the extent to which this type of ribosome modification occurs among other antibiotic producers, especially that of MLS where it was already found that the induction of the resistance to MLS antibiotics is bound to the occurrence of mono- or dimethyladenine in the 23S RNA.[44]

There are remarkable findings about the induction of adenine monomethylation in the 23S rRNA by exogenously added lincomycin in the lincomycin producer *S. lincolnensis* and the possibility of preparing mutants of streptomycetes constitutive in their resistance by a selection with an MLS antibiotic lacking the inductive activity.[44,148]

A definitive answer to what characteristics of the organisms producing antibiotics are responsible for effective resistance mechanisms in vivo can best be obtained after the isolation of the resistance determinants using DNA cloning technology followed by biochemical and genetic analyses of the resistant clones. This approach was employed by Thompson et al.,[149,150] who used a biochemical analysis of the resistant clones of *S. lividans* to demonstrate the presence of enzymes that were originally characterized as the determinants of resistance in the production organisms.

A clone of *S. lividans* TC14 contained 23S rRNA ribose methylase indistinguishable from that of the thiostrepton resistance in *S. azureus* and its ribosomes did not bind thiostrepton.[150] Clones of *S. lividans* TC33 and TC61 possessed the MLS phenotype in vivo; the ribosomes were resistant in vitro to the macrolides carbomycin and niddamycin as well as to lincomycin. Cell-free extracts from the TC33 and TC61 clones mediated the methylation of rRNA from the parent strain of *S. lividans* M180, but not rRNA from the resistant *S. erythreus;* in addition, highly purified 23S rRNA adenine methylase of *S. erythreus* methylated only rRNA from *S. lividans* M180 and not that from the clones TC33 and TC61. The stoichiometry of the methylation by the RNA methylase cloned into *S. lividans* as well as its product were identical with those from the original methylation of *S. erythreus.*[150] Detailed restriction endonuclease cleavage maps of DNA fragments carrying the genes for the ribosomal methylases responsible for the resistance to thiostrepton from *S. azureus* and to MLS antibiotics from *S. erythreus* were constructed and the genes for the resistance were localized in relation to these maps.[151]

B. Complex Protection of the Target Site

More complex mechanisms of target site protection from autotoxic products have been found in other producers of antibiotics acting on proteosynthesis, such as tetracyclines, actinomycins, and aminocyclitols. Apparently, resistance can be ensured at more levels.

The understanding of the mechanisms and their mutual relationships is as yet far from complete.

1. Tetracyclines

Tetracycline antibiotics inhibit proteosynthesis by interfering with the binding of aminoacyl tRNA on the bacterial ribosome acceptor site.[152]

Mikulík et al.[153] compared the proteosynthesis sensitivity to tetracycline in the production and nonproduction strains of *S. aureofaciens*. Addition of 500 μg of tetracycline per milliliter to a 4-hr culture of the production strain resulted in a 37% inhibition of leucine incorporation and a complete inhibition of antibiotic production. Proteosynthesis in the nonproduction strain was completely inhibited under identical experimental conditions. Addition of tetracycline (500 μg/mℓ) to the 16-hr culture of the production strain no longer inhibited the antibiotic production.[17] A considerable difference in the effect of chlortetracycline on RNA synthesis in the low- and high-production strains of *S. aureofaciens* was observed.[20] Exogenous chlortetracycline significantly reduced RNA synthesis in both strains studied only in the growth phase, but not during chlortetracycline biosynthesis. Ultrastructural changes were observed in the producing mycelium of *S. aureofaciens* whose cytoplasm contained a large number of ribosomal aggregates.[154]

Ribosomes and the ribosomal subunits of *S. aureofaciens* bind tetracycline in a way analogous to that of other prokaryotic organisms; the amount of tetracycline bound to the 30S subunits was substantially higher than that bound to the 50S subunits.[153] Mikulík and Vaněk[19] presumed that tetracycline inhibited the binding of the initiation factor IF3 by way of its competition with the RNA binding site where both IF3 and tetracycline interact with mRNA.

The higher resistance of the production strain to tetracycline is apparently associated with a permeability modification. This fact was demonstrated in an experiment monitoring the uptake of ^{3}H-tetracycline by the cells of the high- and low-producing strains of *S. aureofaciens*. Four-hour mycelium of the low-producing strain was found accumulating 23% more of the radioactive tetracycline than that of the high-producing strain.[153] Analogous results were obtained in *S. rimosus* producing oxytetracycline.[155]

In vitro proteosynthesis in a system containing the S30 fraction with endogenous mRNA from a 10-hr culture of the production strain is inhibited by 50% in the presence of 50 μg of tetracycline per milliliter.[153] This implies a higher resistance of the proteosynthetic system of the producing organism than that of sensitive bacteria, but the conducted experiments have not provided an answer concerning the development of a more effective permeability barrier and a ribosomal resistance mechanism in the course of continued cultivation when the production reaches 2000 μg/mℓ.

However, recent results of Mikulík et al.[34] have revealed a much higher sensitivity to tetracycline of ribosomes from both *S. aureofaciens* in which the antibiotic production was suppressed by cultivation conditions, and from the same strain producing tetracycline in comparison with *E. coli* ribosomes during phenylalanine tRNA binding as well as in poly(U)-translation. In addition, there was no difference found between the ribosomal systems of both types of *S. aureofaciens* mycelium in the activity of poly(U)-translation.[34] This would not seem to indicate the existence of a ribosomal resistance mechanism in spite of the above-shown differences in the ribosomal proteins from the producing cells and those where the production was suppressed by changed cultivation conditions. It is, however, possible that in the course of the ribosomal in vitro system preparation, the ribosomal-resistant character was not preserved, being, e.g., removed by washing. No inactivation or modification of the antibiotic by the producing culture of *S. aureofaciens* was observed.[153]

The possibility that the antibiotic target site could be protected only by an effective product excretion out of the cells and by a permeability barrier in the tetracycline-synthesizing microorganisms seems to be less probable. In a number of bacteria resistant to exogenous tetracycline, there are several pieces of evidence indicating the presence of proteins mediating resistance other than just at the transport level.[156-158] A study of the resistance to oxytetracycline in *S. rimosus* also points to the existence of more mechanisms, probably of the plasmid-determined resistance.[155] Its inducible character at subinhibition concentrations of oxytetracycline was demonstrated in experiments involving both growing cultures and proteosynthesis inhibition by oxytetracycline following a preincubation.

2. Actinomycins

As an example of antibiotics, inhibitors at the transcription level, we shall mention the actinomycins which are believed to inhibit DNA-dependent RNA synthesis by an intercalation into DNA.[159] The inhibition effect of actinomycin on the producing organism *S. antibioticus* was investigated by Yoshida et al.[160] They found its effect on amino acid incorporation changing with the age of the culture; the younger mycelium is far more sensitive to actinomycin than are older cells that produce the antibiotic. This was also confirmed by Marshall et al.,[161] who observed a rather quick development of the resistance accompanied by a reduction in the ability of the organism to take up exogenous actinomycin at the time of the synthesis of the enzyme phenoxazinone synthase, but still before the antibiotic synthesis takes place. At the same time, the organism did not lose its sensitivity to chloramphenicol and puromycin.

A correlation of the resistance to actinomycin and the ability to produce this antibiotic in a variety of *S. parvulus* isolates treated with acriflavine was observed.[162] It is interesting that the nonproducing isolates, which lost the resistance, acquired it again after being incubated with 10 μg of actinomycin per milliliter.

The mycelium of *S. parvulus,* synthesizing the antibiotic, is resistant to 5 to 100 μg of actinomycin D per milliliter, while its synthesis by protoplasts prepared from such mycelium was inhibited by approximately 80% in the presence of more than 50 μg of actinomycin D per milliliter.[163] The probable reason for this difference is the loss or a modification of the protection system in the course of protoplast preparation. This system is likely to be localized within the membrane or the cell wall or in between the two. However, the uptake of actinomycin D can be suppressed even in protoplasts by a preincubation of the mycelium of *S. parvulus* with 10 μg of actinomycin D per milliliter.[162] In such instances, protoplasts could become a valuable means for studying the development of resistance to antibiotics mediated via permeability modifications.

Jones[51] attempted to characterize the causes of the changes in the translation apparatus in conjunction with actinomycin production in *S. antibioticus*. tRNA and ribosomes from the producing culture have a reduced ability to function in aminoacylation and translation when compared with the corresponding components from younger nonproducing cells.[29] Specific changes in the pattern of isoaccepting tRNAs were also observed.[164] It was not clear whether these changes were associated with actinomycin production or merely with the age of the cells. In view of the fact that the inhibition of the antibiotic biosynthesis by α-methyl-DL-tryptophan reversed the changes in tRNA normally observed in the control culture, it was suggested that specific tRNAs could somehow be involved in the synthesis of actinomycin. On the other hand, changes in the ribosome activity depend on the age of the culture and cannot be reversed by the inhibition of antibiotic formation. The reduction in RNA synthesis is apparently connected with the production of specific proteins, the transcription inhibitors, which were partially purified from *S. antibioticus* and may not have been directly dependent on the accumulation of actinomycin in the culture.[164]

Apart from the already mentioned protection at the level of the permeability barrier for actinomycin, it appears that *S. antibioticus* makes use of proteins capable of binding actinomycin and occurring in higher amounts during the production phase and which apparently reduce the intracellular content of the free antibiotic.[164]

In addition, in vitro transcription systems prepared from *S. antibioticus* are less sensitive to actinomycin than those from *E. coli.*[164] Highly purified (400 times) RNA polymerase from *S. antibioticus*[48] lost its ability to catalyze transcription resistant to actinomycin, which was observed with crude cell extracts and the partially purified enzyme.[164] It is inferred that the system components responsible for this property were removed during purification, which has also been confirmed by the finding of a gradual increase in the transcription sensitivity to actinomycin with individual purification steps. Thus, the resistance is not an intrinsic property of RNA polymerase itself. Jones[164] has ascribed this to the ability of DNA and RNA polymerase complex to synthesize RNA even in the presence of bound actinomycin.

3. Aminocyclitols

Another important group of antibiotics primarily inhibiting proteosynthesis are the aminocyclitols.[165,166] However, the individual members of this group differ from one another in their ways of action; this is obvious from their different binding sites on the ribosomes and from the existence of resistant mutants of bacteria with altered target sites in which the broad cross-resistance pattern has not been found.[166]

Cella and Vining[21,167] investigated streptomycin action and the type of resistance to this antibiotic in the producing organism *S. griseus.* They found that the cultures of *S. griseus* growing under conditions not supporting antibiotic production were sensitive to streptomycin.[167] Growth cessation was accompanied by proteosynthesis inhibition and an accumulation of the stuck 70S ribosomes. Although the effect of the antibiotic was irreversible and a great majority of the cells were dying, a certain lag period was followed by a new growth of the culture. The reason was the selection of resistant variants from the population. As long as the culture of *S. griseus* grew under the conditions favoring antibiotic biosynthesis, the sensitivity to streptomycin was noted during the growth phase only. With the onset of biosynthesis, the culture was becoming resistant.[21] This was also confirmed by a study on the inhibition of ^{14}C-leucine incorporation in vivo by streptomycin. Measurements of streptomycin uptake by the cells provided evidence in favor of membrane permeability participation in the resistance. However, a certain role was also ascribed to the inactivation of streptomycin by its phosphorylation.

Teraoka and Tanaka[147] observed that the ribosomes from a streptomycin-producing strain of *S. griseus* had a low affinity for dihydrostreptomycin, which could be a factor of resistance. However, these authors also pointed out the possibility of an interference of the present ATP-dependent enzyme inactivating streptomycin. The sensitivity to streptomycin of the ribosomes from the streptomycin-producing *S. griseus* was investigated in systems synthesizing proteins in vitro after having eliminated the undesirable effects of proteolytic enzymes.[168,169] Ribosomes from a 16-hr culture of the production strain of *S. griseus* exhibited the same sensitivity to streptomycin as those from the nonproduction strains or from *E. coli.*[168] The system used in this experiment contained ribosomes of the production strain and a post-ribosomal supernatant from the nonproduction variant, while the streptomycin phosphorylating activity was entirely eliminated from the used ribosomes by washing. Sugiyama et al.[170] observed the sensitivity of proteosynthesis also in a system from a stationary-phase mycelium.

In the course of further detailed study, the difficulties in defining the contributions of a lower ribosome affinity, a streptomycin-modifying enzyme, and a reduced permeability to streptomycin in the overall self-protection of *S. griseus* were demonstrated.[170-174] Of these contributions, the authors consider streptomycin 6-phosphotranferase as the most important.[169,174] Hotta et al.[175] presented the results of their investigation of the effects of kanamycin, neomycin, and streptomycin on in vitro proteosynthesis in extracts from the strains producing these aminocyclitol antibiotics *S. kanamyceticus, S. fradiae,* and *S. griseus,* respectively). In order to characterize the present resistance mechanisms, they used in vitro systems reconstituted by a reciprocal exchange between ribosomes and S150 fractions from the producing (resistant) organism and from another organism sensitive to the antibiotic. They confirmed the sensitivity of the ribosomes from all of the strains toward their own antibiotic and a reduction of the inhibition effects of the antibiotics on proteosynthesis in vitro in the presence of the corresponding inactivating enzymes. In addition, *S. fradiae* and *S. kanamyceticus* were highly resistant to aminocyclitols other than those which they produce. Inactivating enzymes were shown to play a major part in this resistance. Nevertheless, the resistance of *S. kanamyceticus* to streptomycin appears to be at the ribosome level. Piwowarski and Shaw[176] also pointed out the importance of the inactivation enzymes in the resistance to streptomycin in *S. bikiniensis*. In a recent study, Ono et al.[177] concluded that the activity of (hydroxy)streptomycin phosphotransferase is the main cause of the resistance of the hydroxystreptomycin-producing *S. glaucescens* to aminoglycosides of the streptomycin type. One possible explanation for the simultaneous loss of the antibiotic production and a resistance to it in mutants is that phosphotransferase is also involved in the biosynthesis of hydroxystreptomycin.

On the other hand, Ogawa and Perlman[179] studied the occurrence of the streptomycin phosphorylating enzyme in the streptomycin-producing actinophage-resistant variant obtained from the nonproduction strain of *S. griseus*. They found that this producing culture had no ability to inactivate the antibiotic by phosphorylation. Yamamoto et al.[180] discovered that the istamycin-producing *S. tenjimariensis* has no inactivating enzymes, although its ribosomes are resistant to istamycins. The characteristic resistance pattern of this organism to a variety of other aminocyclitol antibiotics depends upon its ribosomal resistance.[53] Among the aminoglycoside producers, *S. tenebrarius* producing nebramycin exhibited the widest range of resistance: it was highly resistant to all aminocyclitols tested.[181] The results of the investigation of its resistance demonstrated the presence of both resistant ribosomes and inactivating enzymes: aminoglycoside *N*-acetyltransferase(s) and aminoglycoside 6-phosphotransferase. The same pattern and mechanism of resistance proved to exist in strains having the same antibiotic productivity and growth appearance.[182]

S. fradiae, a neomycin producer, contains ribosomes sensitive to neomycin.[175,183] Neomycin-modifying enzymes 3′-*O*-phosphotransferase and 3-*N*-acetyltransferase were detected in this organism.[184] Thompson et al.[150] analyzed biochemically the clones of *S. lividans* containing the determinants of neomycin resistance from *S. fradiae,* and found that neither phosphotransferase activity present in the clone TC8, nor acetyltransferase activity in the clone TC7 could by themselves ensure a high level of resistance to neomycin. Only when both enzymes were introduced into the same strain, clone TC65, was the obtained resistance level comparable with the original level in *S. fradiae*. When both neomycin-modifying enzymes are present in the same cells, a synergistic effect on the resistance is observed. At the same time, gene expression for phosphotransferase in *S. lividans* was 50 times higher than in the original antibiotic-producing organism.

Another example of recent findings on the regulation of the synthesis of proteins responsible for self-protection is the observation of an induction of resistance in the A-factor-negative mutants of *S. griseus* and *S. bikiniensis* following the addition of the A factor (see Section IV), which, apart from the enhanced streptomycin production, also initiated the synthesis of streptomycin phosphotransferase.[185]

Increased formation of modifying enzymes leading to higher resistance levels enables a higher antibiotic biosynthesis in cases where the resistance is rate limiting.[166,186]

The origin of the determinants of resistance to antibiotics coded for by extrachromosomal elements in clinical isolates[187] has been ascribed to the producers of the antibiotics. Walker and Walker[188] and Walker and Skorvaga[189] suggested that the enzymes modifying streptomycin in *S. griseus* and *S. bikiniensis* could be related to the enzymes produced by bacteria containing R plasmids. Benveniste and Davies[190] found in *S. kanamyceticus* an enzyme acetylating the 6′ amino group of kanamycin A and B, gentamycin C_{1a} and neomycin, and another enzyme in *Streptomyces spectabilis* which acetylates the 2′ amino group of gentamycin C_{1a}. These enzymes catalyze reactions that are identical with those catalyzed by enzymes from resistant clinically isolated bacteria.[184] Benveniste and Davies[190] pointed out a possible evolutionary relationship between enzymes from actinomycetes and R^+ bacteria. The question of the origin of the genes of antibiotic resistance can be answered after more information on the nucleotide sequences of these genes becomes available.[191] The idea of a gradual development of resistance in antibiotic producers proceeding along with the evolution of antibiotic production[192] is consistent with the findings of complex protection of target site from the toxic drug effects secured at a number of levels.

REFERENCES

1. **Demain, A. L.,** Mutation and the production of secondary metabolites, *Adv. Appl. Microbiol.,* 16, 177, 1973.
2. **Hošťálek, Z.,** Regulation of gene expression in secondary biosynthesis, in *Regulation of Secondary Product and Plant Hormone Metabolism,* Luckner, M. and Schreiber, K., Eds., Pergamon Press, New York, 1979, 111.
3. **Martin, J. F. and Demain, A. L.,** Control of antibiotic biosynthesis, *Microbiol. Rev.,* 44, 230, 1980.
4. **Morihara, K., Oka, T., and Tsuzuki, H.,** Multiple proteolytic enzymes of *Streptomyces fradiae* production, isolation, and preliminary characterization, *Biochim. Biophys. Acta,* 139, 382, 1967.
5. **Morihara, K.,** Comparative specificity of microbial proteinases, *Adv. Enzymol.,* 41, 179, 1974.
6. **Laluce, C. and Molinari, R.,** Fractionation of the proteolytic and amylolytic complex enzyme system of *Streptomyces aureofaciens* and some properties of fractions, *Biotechnol. Bioeng.,* 21, 915, 1979.
7. **Awad, W. M., Soto, A. R., Siegel, S., Skiba, W. E., Bernstrom, G. G., and Ochoa, M. S.,** The proteolytic enzymes of the K-1 strain of *Streptomyces griseus* obtained from a commercial preparation (pronase). I. Purification of four serine endopeptidases, *J. Biol. Chem.,* 247, 4144, 1972.
8. **Pokorny, M., Vitale L., Turk, V., Renko, M., and Žuvanić, J.,** *Streptomyces rimosus* extracellular proteases. I. Characterization and evaluation of various crude preparations, *Eur. J. Appl. Microbiol. Biotechnol.,* 8, 81, 1979.
9. **Renko, M., Pokorny, M., Vitale, L., and Turk, V.,** *Streptomyces rimosus* extracellular proteases. II. Isolation and characterization of serine alkaline proteinase, *Eur. J. Appl. Microbiol. Biotechnol.,* 11, 166, 1981.
10. **Narahashi, Y. and Yoda, K.,** Purification and some properties of a new metallo carboxypeptidase of *Streptomyces griseus* K-1, *J. Biochem. (Tokyo),* 86, 683, 1979.
11. **Olafson, R. W. and Smillie, L. B.,** Enzymic and physicochemical properties of *Steptomyces griseus* trypsin, *Biochemistry,* 14, 1161, 1975.
12. **Radzhabov, L. R., Shibnev, V. A., and Rassulin, Y. A.,** On esterase and peptidase activity of elastolytic enzyme from *Actinomyces rimosus (in Russian), Khim. Prir. Soed., SSSR,* 5, 693, 1977.
13. **Ginther, Ch. L.,** Sporulation and the production of serine protease and cephamycin C by *Streptomyces lactamdurans, Antimicrob. Agents Chemother.,* 15, 522, 1979.
14. **Bradley, S. G.,** Posttranscriptional regulation of metabolic processes in *Actinomycetes,* in *Actinomycetes,* Zbl. Bakt. Suppl. No. 11, Schaall, K. P. and Pulverer, G., Eds., Gustav Fischer Verlag, New York, 1981, 367.
15. **Katz, E. and Weissbach, H.,** Incorporation of ^{14}C-labeled amino acids into actinomycin and protein by *Streptomyces antibioticus, J. Biol. Chem.,* 238, 666, 1963.
16. **Jones, G. H. and Weissbach, H.,** RNA metabolism in *Streptomyces antibioticus*: effect of 5-fluorouracil on the appearance of phenoxazinone synthetase, *Arch. Biochem. Biophys.,* 137, 558, 1970.
17. **Mikulík, K., Karnetová, J., Quyen, N., Blumauerová, M., Komersová, I., and Vaněk, Z.,** Interaction of tetracycline with protein synthesizing system of *Streptomyces aureofaciens, J. Antibiot.,* 24, 801, 1971.
18. **Běhal, V., Vaněk, Z., Hošťálek, Z., and Ramadan, A.,** Synthesis and degradation of proteins and DNA in *Streptomyces aureofaciens, Folia Microbiol.,* 24, 211, 1979.
19. **Mikulík, K. and Vaněk, Z.,** Interrelationship between primary and secondary metabolism in actinomycetes, in *Regulation of Secondary Product and Plant Hormone Metabolism,* Luckner, M. and Schreiber, K., Eds., Pergamon Press, New York, 1979, 199.

20. **Šimúth, J., Hudec, J., Chau, H. T., Dányi, O., and Zelinka, J.**, The synthesis of highly phosphorylated nucleotides, RNA and protein by *Streptomyces aureofaciens, J. Antibiot.*, 32, 53, 1979.
21. **Cella, R. and Vining, L. C.**, Resistance to streptomycin in a producing strain of *Streptomyces griseus, Can. J. Microbiol.*, 21, 463, 1975.
22. **Liras, P., Villanueva, J. R., and Martín, J. F.**, Sequential expression of macromolecule biosynthesis and candicidin formation in *Streptomyces griseus, J. Gen. Microbiol.*, 102, 269, 1977.
23. **Gersch, D.**, Dependence of macromolecule synthesis on cyclic AMP in antibiotic producing strains of *Streptomyces hygroscopicus, FEMS Microbiol. Lett.*, 6, 343, 1979.
24. **Šťastná, J. and Mikulík, K.**, Role of highly phosphorylated nucleotides and antibiotics in the development of *Streptomycetes*, in *Actinomycetes*, Zbl. Bakt. Suppl. No. 11, Schaal, K. P. and Pulverer, G., Eds., Gustav Fischer Verlag, New York, 1981, 481.
25. **Mikulík, K., Janda, I., Mašková, H., Šťastná, J., and Jiráňová, A.**, Macromolecular synthesis accompanying the transition from spores to vegatative forms of *Streptomyces granaticolor, Folia Microbiol.*, 22, 252, 1977.
26. **Hardisson, C., Salas, J. A., Guijarro, J. A., and Suarez, J. E.**, Macromolecular synthesis during the germination of *Streptomyces* spores in a chemically defined medium, *FEMS Microbiol. Lett.*, 7, 233, 1980.
27. **Hardisson, C., Manzanal, M. B., Salas, J. A., and Suarez, J. E.**, Fine structure, physiology and biochemistry of arthrospore germination in *Streptomyces antibioticus, J. Gen. Microbiol.*, 105, 203, 1978.
28. **Guijarro, J. A., Suarez, J. E., Salas, J., and Hardisson, C.**, Characterization of RNA species synthesized during germination and vegetative growth of *Streptomyces antibioticus, FEMS Microbiol. Lett.*, 14, 205, 1982.
29. **Jones, G. H.**, Macromolecular synthesis in *Streptomyces antibioticus: in vitro* systems for aminoacylation and translation from young and old cells, *J. Bacteriol.*, 124, 364, 1975.
30. **Mikulík, K., Janda, I., Řičicová, A., and Vinter, V.**, Differentiation in *Actinomycetes*. I. Ribosomes of vegetative cells and spores of *Streptomyces granaticolor*, in *Spores*, Vol. 6, Gerhardt, P., Costilow, R. N., and Sadoff, H. L., Eds., American Society for Microbiology, Washington, D.C., 1975, 15.
31. **Valu, G. and Szabó, G.**, Characterization of a cell-free polypeptide synthesizing system from *Streptomyces griseus*, in *Actinomycetes*, Zbl. Bakt. Suppl. No. 11, Schaal, K. P. and Pulverer, G., Eds., Gustav Fischer Verlag, New York, 1981, 487.
32. **Janda, I., Mikulík, K., and Weiser, J.**, Ribosomal proteins of *Streptomyces granaticolor, Biochim. Biophys. Acta*, 656, 228, 1981.
33. **Mikulík, K., Janda, I., Weiser, J., and Jiráňová, A.**, Ribosomal proteins of *Streptomyces aureofaciens* producing tetracycline, *Biochim. Biophys. Acta*, 699, 203, 1982.
34. **Mikulík, K., Jiráňová, A., Janda, I., and Weiser, J.**, Susceptibility of ribosomes of the tetracycline producing strain of *Streptomyces aureofaciens* to tetracyclines, *FEBS Lett.*, 152, 125, 1983.
35. **Valu, G., Békési, I., and Szabó, G.**, Ribosomal proteins from vegetative hyphae and from spores of *Streptomyces griseus, Acta Biol. Acad. Sci. Hung.*, 26, 151, 1975.
36. **Vaněk, Z. and Mikulík, K.**, Microbial growth and production of antibiotics, *Folia Microbiol.*, 23, 309, 1978.
37. **Loening, U.E.**, Molecular weights of ribosomal RNA in relation to evolution, *J. Mol. Biol.*, 38, 355, 1968.
38. **Timko, J. and Zelinka, J.**, Ribosomal ribonucleic acids from *Streptomyces griseus, J. Antibiot.*, 30, 644, 1977.
39. **Valu, G. and Szabó, G.**, High molecular weight ribosomal ribonucleic acids from vegetative hyphae and spores of *Streptomyces griseus, Acta Microbiol. Acad. Sci. Hung.*, 25, 229, 1978.
40. **Cundliffe, E.**, Mechanism of resistance to thiostrepton in the producing-organism *Streptomyces azureus, Nature (London)*, 272, 792, 1978.
41. **Cundliffe, E. and Thompson, J.**, Ribose methylation and resistance to thiostrepton, *Nature (London)*, 278, 859, 1979.
42. **Thompson, J. and Cundliffe, E.**, Purification and properties of an RNA methylase produced by *Streptomyces azureus* and involved in resistance to thiostrepton, *J. Gen. Microbiol.*, 124, 291, 1981.
43. **Graham, M. Y. and Weisblum, B.**, 23S ribosomal ribonucleic acid of macrolide-producing streptomycetes contains methylated adenine, *J. Bacteriol.*, 137, 1464, 1979.
44. **Fujisawa, Y. and Weisblum, B.**, A family of r-determinants in *Streptomyces* spp. that specifies inducible resistance to macrolide, lincosamide and streptogramin type B antibiotics, *J. Bacteriol.*, 146, 621, 1981.
45. **Skinner, R. H. and Cundliffe, E.**, Dimethylation of adenine and the resistance of *Streptomyces erythreus* to erythromycin, *J. Gen. Microbiol.*, 128, 2411, 1982.
46. **Weiser, J., Mikulík, K., and Bosch, L.**, Studies on the elongation factor Tu from *Streptomyces aureofaciens* producing tetracycline, *Biochem. Biophys. Res. Commun.*, 99, 16, 1981.
47. **Weiser, J., Mikulík, K., Žižka, Z., Šťastná, J., Janda, I., and Jiráňová, A.**, Isolation and characterization of *Streptomyces aureofaciens* protein-synthesis elongation factor Tu in an aggregated state, *Eur. J. Biochem.*, 129, 127, 1982.

48. **Jones, G. H.,** Purification of RNA polymerase from actinomycin producing and nonproducing cells of *Streptomyces antibioticus, Arch. Biochem. Biophys.,* 198, 195, 1979.
49. **Watanabe, S. and Tanaka, K.,** Effect of rifampicin on *in vitro* RNA synthesis of *Streptomyces mediterranei, Biochem. Biophys. Res. Commun.,* 72, 522, 1976.
50. **Valu, G. and Szabó, G.,** The effect of endogenous proteolytic activity on the *in vitro* ^{14}C-phenylalanine incorporation in *Streptomyces griseus,* in *Nocardia and Streptomyces,* Mordarski, M., Kurylowicz, W., and Jeljaszewicz, J., Eds., Gustav Fischer Verlag, New York, 1978, 409.
51. **Jones, G. H.,** Relationship between changes in the translational apparatus and actinomycin production in *Streptomyces antibioticus, J. Bacteriol.,* 129, 81, 1977.
52. **Thompson, J., and Cundliffe, E.,** Resistance to thiostrepton, siomycin, and sporangiomycin in *Actinomycetes* that produce them, *J. Bacteriol.,* 142, 455, 1980.
53. **Yamamoto, H., Hotta, K., Okami, Y., and Umezawa, H.,** Ribosomal resistance of an istamycin producer *Streptomyces tenjimariensis,* to aminoglycoside antibiotics, *Biochem. Biophys. Res. Commun.,* 100, 1396, 1981.
54. **Martin, M. and Sabater, B.,** Cell-free synthesis of catalytically active fructokinase directed by RNA from *Streptomyces violaceoruber* grown with fructose, *J. Gen. Microbiol.,* 126, 203, 1981.
55. **Jones, G. H.,** Cell-free synthesis of the phenoxazinone synthetase subunit by polyribosomes from *Streptomyces antibioticus, J. Biol. Chem,.,* 255, 1945, 1980.
56. **Trnovský, J., Šimúth, J., Zelinka, J., and Hanus, L.,** Cell-free proteosynthetic system from *Streptomyces aureofaciens.* II. Endogenous mRNA translation, *Biológia (Bratislava),* 37, 369, 1982.
57. **Hash, J. H.,** Antibiotics, *Methods Enzymol.,* 43, 427, 1975.
58. **Walker, J. B.,** Pathways of biosynthesis of the guanidinated inositol moieties of streptomycin and bluensomycin, *Methods Enzymol.,* 43, 429, 1975.
59. **Vaněk, Z., Cudlín, J., Blumauerová, M., and Hošťálek, Z.,** How many genes are required for the synthesis of chlortetracycline? *Folia Microbiol.,* 16, 225, 1971.
60. **Lynen, F. and Tada, M.,** Die Biochemische Grundlagen der "Polyacetat-Regel", *Angew. Chem.,* 73, 513, 1961.
61. **Dimroth, P., Walter, H., and Lynen, F.,** Biosynthese von 6-Methylsalicylsäure, *Eur. J. Biochem.,* 13, 98, 1970.
62. **Vogel, G. and Lynen, F.,** 6-Methylsalicylic acid synthetase, *Methods Enzymol.,* 43, 520, 1975.
63. **Bu'Lock, J. D., Hamilton, D., Hulme, M. A., Powell, A. J., Smalley, H. M., Shepherd, D., and Smith, G. N.,** Metabolic development and secondary biosynthesis in *Penicillium urticae, Can. J. Microbiol.,* 11, 765, 1965.
64. **Bu'Lock, J. D., Shepherd, D., and Winstanley, D. J.,** Regulation of 6-methylsalicylate and patulin synthesis in *Penicillium urticae, Can. J. Microbiol.,* 15, 279, 1969.
65. **Forrester, P. I. and Gaucher, G. M.,** Conversion of 6-methylsalicylic acid into patulin by *Penicillium urticae, Biochemistry,* 11, 1102, 1972.
66. **Forrester, P. I. and Gaucher, G. M.,** m-Hydroxbenzyl alcohol dehydrogenase from *Penicillium urticae, Biochemistry,* 11, 1108, 1972.
67. **Light, R. J. and Vogel, G.,** 6-Methylsalicylic acid (2,6-cresotic acid) decarboxylase, *Methods Enzymol.,* 43, 530, 1975.
68. **Gaucher, G. M.,** m-Hydroxybenzyl-alcohol dehydrogenase, *Methods Enzymol.,* 43, 540, 1975.
69. **Scott, A. I. and Beadling, L. C.,** Biosynthesis of patulin. Dehydrogenase and dioxygenase enzymes of *Penicillium urticae, Biorg. Chem.,* 3, 281, 1974.
70. **Gaucher, G. M. and Sekiguchi, J.,** New steps in the post-gentisaldehyde portion of the patulin biosynthetic pathway, presented at 5th Int. Ferment. Symp., West Berlin, 1976, 229.
71. **Miller, P. A. and Hash, J. H.,** *S*-adenosylmethionine: dedimethylamino-4-aminoanhydrotetracyline *N*-methyltransferase, *Methods Enzymol.,* 43, 603, 1975.
72. **Běhal, V., Hošťálek, Z., and Vaněk, Z.,** Anhydrotetracycline oxygenase activity and biosynthesis of tetracyclines in *Streptomyces aureofaciens, Biotechnol. Lett.,* 1, 177, 1979.
73. **Běhal, V., Neužil, J., and Hošťálek, Z.,** Effect of tetracycline derivatives and some cations on the activity of anhydrotetracycline oxygenase, *Biotechnol. Lett.,* 5, 537, 1983.
74. **Miller, P. A. and Hash, J. H.,** NADP: tetracycline 5a(11a) dehydrogenase, *Methods Enzymol.,* 43, 606, 1975.
75. **Kominek, L. A. and Meyer, H. F.,** Novobiocic acid synthetase, *Methods Enzymol.,* 43, 502, 1975.
76. **Rafalski, A. and Raczynska-Bojanowska K.,** Synthesis of malonate and methylmalonate and the formation of polyene antibiotics, *Acta Biochim. Polon.,* 19, 71, 1972.
77. **Gräfe, U., Bocker, H., and Thrum, H.,** Precursorbildung und Biosynthese des Makrolidantibioticums A 6599, (Turimycin) durch *Streptomyces hygroscopicus* JA 6599, *Z. Allg. Mikrobiol.,* 15, 249, 1975.
78. **Běhal, V. and Vaněk, Z.,** Regulation of biosynthesis of secondary metabolite. XII. Acetyl-CoA carboxylase in *Streptomyces aureofaciens, Folia Microbiol.,* 15, 354, 1970.

79. **Běhal, V., Jechová, V., Vaněk, Z., and Hošťálek, Z.,** Alternate pathways of malonyl-CoA formation in *Streptomyces aureofaciens, Phytochemistry,* 16, 347, 1977.
80. **Gatenbeck, S. and Mahlén, A. K.,** An alternate route for the formation of malonate in *Penicillium islandicum, Acta Chem. Scand.,* 22, 1696, 1968.
81. **Venkitasubramanian, T. A., Bhatnagar, R. K., Saraswathy, S., Rao, V. B., and Sivaswami, J.,** Intermediary metabolism in *Aspergillus parasiticus* in relation to aflatoxin biosynthesis, in *Overproduction of Microbial Products,* Krumphanzl, V., Sikyta, B., and Vaněk, Z., Eds., Academic Press, New York, 1982, 153.
82. **Krupinski, V. M., Robbers, J. E., and Floss, H. G.,** Physiological study of ergot: induction of alkaloid synthesis by tryptophan at the enzymatic level, *J. Bacteriol.,* 125, 158, 1976.
83. **Drew, S. W. and Demain, A. L.,** Methionine control of cephalosporin C formation, *Biotechnol. Bioeng.,* 15, 743, 1973.
84. **Rogers, T. O. and Birnbaum, J.,** Biosynthesis of fosfomycin by *Streptomyces fradiae, Antimicrob. Agents Chemother.,* 5, 121, 1974.
85. **Khokhlov, A. S., Tovarova, I. I., Borisova, L. N., Pliner, S. A., Shevchenko, L. A., Kornitskaya, E. J., Ivkina, N. S., and Rapoport, I. A.,** The A-factor responsible for the biosynthesis of streptomycin in mutant strains of *Actinomyces streptomycini* (in Russian), *Dokl. Akad. Nauk SSSR,* 177, 232, 1967.
86. **Yanagimoto, M. and Terui, G.,** Physiological studies on staphylomycin production. II. Formation of substance effective in inducing staphylomycin production, *J. Ferment. Technol.,* 49, 611, 1971.
87. **Khokhlov, A. S. and Tovarova, I. I.,** Studies on the streptomycin biosynthesis, *Post. Hig. Med. Dośw.,* 26, 469, 1972.
88. **Khokhlov, A. S., Tovarova, I. I., and Anisova, L. N.,** Regulators of streptomycin biosynthesis and development of *Actinomyces streptomycini,* in *Secondary Metabolism and Coevolution,* Luckner, M., Mothes, K., and Nover, L., Eds., Deutsche Akademie der Naturforscher Leopoldina, Halle, 1976, 289.
89. **Khokhlov, A. S., Anisova, L. N., Tovarova, I. I., Kleiner, E. M., Kovalenko, I. V., Krasilnikova, O. I., Kornitskaya, E. Ya., and Pliner, S. A.,** Effect of A-factor on the growth of asporogenous mutants of *Streptomyces griseus,* not producing this factor, *Z. Allg. Mikrobiol.,* 13, 647, 1973.
90. **Kleiner, E. M., Pliner, S. A., Soifer, V. S., Onoprienko, V. V., Balashova, T. A., Rozynov, B. V., and Khokhlov, A. S.,** Structure of the A-factor, a bioregulator from *Streptomyces griseus* (in Russian), *Biorg. Khim.,* 2, 1142, 1976.
91. **Kleiner, E. M., Onoprienko, V. V., Pliner, S. A., Soifer, V. S., and Khokhlov, A. S.,** The synthesis of racemic A-factor — a biological regulator from *Streptomyces griseus (Actinomyces streptomycini)* (in Russian), *Bioorg. Khim.,* 3, 424, 1977.
92. **Hara, O. and Beppu, T.,** Mutants blocked in streptomycin production in *Streptomyces griseus* — the role of A-factor, *J. Antibiot.,* 35, 349, 1982.
93. **Eritt, I., Gräfe, U., and Fleck, W.,** A screening method for autoregulators of anthracycline-producing streptomycetes, *Z. Allg. Mikrobiol.,* 22, 91, 1982.
94. **Gräfe, U., Reinhardt, G., Schade, W., Krebs, D., Eritt, I., and Fleck, W. F.,** Isolation and structure of novel autoregulators from *Streptomyces griseus, J. Antibiot.,* 35, 609, 1982.
95. **Gräfe, U., Schade, W., Eritt, I., and Fleck, W.F.,** A new inducer of anthracycline biosynthesis from *Streptomyces viridochromogenes, J. Antiobiot.,* 35, 1722, 1982.
96. **Demain, A.,** Regulatory mechanisms and the industrial production of microbial metabolites, *Lloydia,* 31, 395, 1968.
97. **Hošťálek, Z.,** Catabolite regulation of antibiotic biosynthesis, *Folia Microbiol.,* 25, 445, 1980.
98. **Majumdar, M. K. and Majumdar, S. K.,** Synthesis of neomycin by washed mycelium of *Streptomyces fradiae* and some physiological considerations, *Folia Microbiol.,* 16, 285, 1971.
99. **Kimura, A.,** Biochemical studies on *Streptomyces sioyaensis.* II. Mechanism of the inhibitory effect of glucose on siomycin formation, *Agric. Biol. Chem.,* 31, 845, 1967.
100. **Aharonowitz, Y. and Demain, A. L.,** Carbon catabolite regulation of cephalosporin production in *Streptomyces clavuligerus, Antimicrob. Agents Chemother.,* 14, 159, 1978.
101. **Kennel, Y. M. and Demain, A. L.,** Effect of carbon sources on beta-lactam antibiotic formation by *Cephalosporium acremonium, Exp. Mycol.,* 2, 234, 1978.
102. **Gallo, M. and Katz, E.,** Regulation of secondary metabolite biosynthesis: catabolite repression of phenoxazinone synthase and actinomycin formation by glucose, *J. Bacteriol.,* 109, 659, 1972.
103. **Sankaran, L. and Pogell, B. M.,** Biosynthesis of puromycin in *Streptomyces alboniger:* regulation and properties of *O*-demethylpuromycin *O*-methyltransferase, *Antimicrob. Agents Chemother.,* 8, 721, 1975.
104. **Demain, A. L.,** How do antibiotic-producing microorganisms avoid suicide? *Ann. N.Y. Acad. Sci.,* 235, 601, 1974.
105. **Jones, A. and Westlake, D. W. S.,** Regulation of chloramphenicol synthesis in *Streptomyces* sp. 3022a. Properties of arylamine synthetase, an enzyme involved in antibiotic biosynthesis, *Can. J. Microbiol.,* 20, 1599, 1974.

106. **Matteo, C. C., Glade, M., Tanaka, A., Piret, J. M., and Demain, A. L.,** Microbiological studies on the formation of gramicidin S synthetases, *Biotechnol. Bioeng.*, 17, 129, 1975.
107. **Kominek, L. A.,** Biosynthesis of novobiocin by *Streptomyces niveus, Antimicrob. Agents Chemother.*, 1, 123, 1972.
108. **Heinstein, P. and Floss, H. G.,** Feedback inhibition and end product repression in secondary metabolism, in *Secondary Metabolism and Coevolution,* Luckner, M., Mothes, K., and Nover, L., Eds., Deutsche Akademie der Naturforscher Leopoldina, Halle, 1976, 299.
109. **Walker, M. S. and Walker, J. B.,** Streptomycin biosynthesis. Separation and substrate specificities of phosphatases acting on guanidino-deoxy-*scyllo*-inositol phosphate and streptomycin-(*streptidino*) phosphate, *J. Biol. Chem.*, 246, 7034, 1971.
110. **Martín, J. F., Liras, P., and Demain, A. L.,** Inhibition by phosphate of the activity of candicidin synthases, *FEMS Microbiol. Lett.*, 2, 173, 1977.
111. **Běhal V., Grégrová-Prušáková, J., and Hošťálek, Z.,** Effect of inorganic phosphate and benzyl thiocyanate on the activity of anhydrotetracycline oxygenase in *Streptomyces aureofaciens, Folia Microbiol.*, 27, 102, 1982.
112. **Madry, N. and Pape, H.,** Formation of secondary metabolism enzymes in the tylosin producer *Streptomyces* T 59-235, *Arch. Microbiol.*, 131, 170, 1982.
113. **Seno, E. T. and Baltz, R. H.,** Properties of *S*-adenosyl-L-methionine:macrocin *O*-methyltransferase in extracts of *Streptomyces fradiae* strains which produce normal or elevated levels of tylosin and in mutants blocked in specific O-methylations, *Antimicrob. Agents Chemother.*, 20, 370, 1981.
114. **Seno, E. T. and Baltz, R. H.,** *S*-adenosyl-L-methionine:macrocin O-methyltransferase activities in a series of *Streptomyces fradiae* mutants that produce different levels of the macrolide antibiotic tylosin, *Antimicrob. Agents Chemother.*, 21, 758, 1982.
115. **Martin, J. F.,** Control of Antibiotic synthesis by phosphate, *Adv. Biochem. Eng.*, 6, 105, 1977.
116. **Hošťálek, Z., Běhal, V., Novotná, J., Erban, V., Čurdová, E., and Jechová, V.,** Regulation of expression of chlortetracycline synthesis in *Streptomyces aureofaciens,* in *Overproduction of Microbial Products,* Krumphanzl, V., Sikyta, B., and Vaněk, Z., Eds., Academic Press, London, 1982, 47.
117. **Čurdová, E., Křemen, A., Vaněk, Z., and Hošťálek, Z.,** Regulation and biosynthesis of secondary metabolites. XVIII. Adenylate level and chlortetracycline production in *Streptomyces aureofaciens, Folia Microbiol.*, 21, 481, 1976.
118. **Vu-Trong, K., Bhuwapathanapun, S., and Gray, P. P.,** Metabolic regulation in tylosin-producing *Streptomyces fradiae*. Regulatory role of adenylate nucleotide pool and enzymes involved in biosynthesis of tylonolide precursors, *Antimicrob. Agents Chemother.*, 17, 519, 1980.
119. **Vu-Trong, K., Bhuwapathanapun, S., and Gray, P. P.,** Metabolic regulation in tylosin producing *Streptomyces fradiae:* phosphate control of tylosin biosynthesis, *Antimicrob. Agents Chemother.*, 19, 209, 1981.
120. **Drew, S. W. and Demain, A. L.,** Effect of primary metabolites on secondary metabolism, *Annu. Rev. Microbiol.*, 31, 343, 1977.
121. **Hošťálek, Z., Tintěrová, M., Jechová, V., Blumauerová, M., Suchý, J., and Vaněk, Z.,** Regulation of biosynthesis of secondary metabolites. I. Biosynthesis of chlortetracycline and tricarboxylic acid cycle activity, *Biotechnol. Bioeng.*, 11, 539, 1969.
122. **Lowe, D. A. and Westlake, D. W. S.,** Regulation of chloramphenicol synthesis in *Streptomyces* sp. 3022a. Branch-point enzymes of the shikimic acid pathway, *Can. J. Biochem.*, 50, 1064, 1972.
123. **Jones, A. and Westlake, D. W. S.,** Regulation of chloramphenicol synthesis in *Streptomyces* sp. 3022a. Properties of arylamine synthetase, an enzyme involved in antibiotic biosynthesis, *Can. J. Microbiol.*, 20, 1599, 1974.
124. **Lowe, D. A. and Westlake, D. W. S.,** Regulation of chloramphenicol synthesis in *Streptomyces* sp. 3022a. 3-deoxy-D-*arabino*-heptulosonate-7-phosphate synthetase, *Can. J. Biochem.*, 49, 448, 1971.
125. **Mendelovitz, S. and Aharonowitz, Y.,** Regulation of cephamycin C synthesis, aspartokinase, dihydrodipicolinic acid synthetase, and homoserine dehydrogenase by aspartic acid family amino acids in *Streptomyces clavuligerus, Antimicrob. Agents Chemother.*, 21, 74, 1982.
126. **Dulaney, E. L.,** Observations on *Streptomyces griseus,* II. Nitrogen sources for growth and streptomycin production, *J. Bacteriol.*, 56, 305, 1948.
127. **Smith, R. L., Bungay, H. R., and Pittenger, R. C.,** Growth-biosynthesis relationships in erythromycin fermentation, *Appl. Microbiol.*, 10, 293, 1962.
128. **Westlake, D. W. S., Sala, F., McGrath, R., and Vining, L. C.,** Influence of nitrogen source on formation of chloramphenicol in cultures of *Streptomyces* sp. 3022a, *Can. J. Microbiol.*, 14, 587, 1968.
129. **Gräfe, U., Bocker, H., Reinhardt, G., and Thrum, H.,** Regulative Beeinflussung der Nourseothricin-biosynthese durch o-Aminobenzosäure in Kulturen des *Streptomyces noursei* JA 3890b, *Z. Allg. Mikrobiol.*, 14, 659, 1974.

130. **Gräfe, U., Bocker, H., and Thrum, H.,** Regulative influence of *o*-aminobenzoic acid on the biosynthesis of nourseothricin in cultures of *Streptomyces noursei* JA 3890b. II. Regulation of glutamine synthetase and the role of glutamine synthetase/glutamate synthase pathway, *Z. Allg. Mikrobiol.*, 17, 201, 1977.
131. **Surikova, E. I., Novikova, L. M., and Gerasimova, T. M.,** Study on regularities of *Actinomyces antibioticus* growth and oleandomycin biosynthesis (in Russian), *Antibiotiki*, 21, 195, 1976.
132. **Aharonowitz, Y. and Demain, A. L.,** Nitrogen nutrition and regulation of cephalosporin production in *Streptomyces clavuligerus, Can. J. Microbiol.*, 25, 61, 1979.
133. **Vining, L. C. and Chatterjee, S.,** Catabolite repression and the control of secondary metabolism, in *Overproduction of Microbial Products,* Krumphanzl, V., Sikyta, B., and Vaněk, Z., Eds., Academic Press, New York, 1982, 35.
134. **Chatterjee, S., Vining, L. C., and Westlake, D. W. S.,** Nutritional requirements for chloramphenicol biosynthesis in *Streptomyces venezuelae, Can. J. Microbiol.*, 29, 247, 1983.
135. **Aharonowitz, Y.,** Nitrogen metabolite regulation of antibiotic biosynthesis, *Annu. Rev. Microbiol.*, 34, 209, 1980.
136. **Gräfe, U.,** Relevance of microbial nitrogen metabolism to production of secondary metabolites, in *Overproduction of Microbial Products,* Krumphanzl, V., Sikyta, B., and Vaněk, Z., Eds., Academic Press, New York, 1982, 63.
137. **Aharonowitz, Y.,** Regulatory interrelationships of nitrogen metabolism and cephalosporin biosynthesis, in *Genetics of Industrial Microorganisms,* Sebek, O. K. and Laskin, A. I., Eds., American Society for Microbiology, Washington, D.C., 1979, 210.
138. **Wax, R., Synder, L., and Kaplan, L.,** Inactivation of glutamine synthetase by ammonia shock in the Gram-positive bacterium *Streptomyces cattleya, Appl. Environ. Microbiol.*, 44, 1004, 1982.
139. **Streicher, S. L. and Tyler, B.,** Regulation of glutamine synthetase activity by adenylylation in the Gram-positive bacterium *Streptomyces cattleya, Proc. Natl. Acad. Sci. U.S.A.*, 78, 229, 1981.
140. **Magasanik, B.,** Genetic control of nitrogen assimilation in bacteria, *Annu. Rev. Genet.*, 16, 135, 1982.
141. **Gale, E. F., Reynolds, P. E., Cundliffe, E:, Richmond, M. H., and Waring, M. J.,** *The Molecular Basis of Antibiotic Action,* (Russian transl.), Mir, Moscow, 1975, chap. 7.
142. **Vining, L. C.,** Antibiotic tolerance in producer organisms, *Adv. Appl. Microbiol.*, 25, 147, 1979.
143. **Gale, E. F., Reynolds, P. E., Cundliffe, E., Richmond, M. H., and Waring, M. J.,** *The Molecular Basis of Antibiotic Action,* (Russian transl.), Mir, Moscow, 1975, 390.
144. **Thompson, J., Cundliffe, E., and Stark, M.,** Binding of thiostrepton to a complex of 23S rRNA with ribosomal protein L11, *Eur. J. Biochem.*, 98, 261, 1979.
145. **Cundliffe, E. and Thompson, J.,** The mode of action of nosiheptide (multhiomycin) and the mechanism of resistance in the producing organism, *J. Gen. Microbiol.*, 126, 185, 1981.
146. **Thompson, J., Cundliffe, E., and Stark, M. J. R.,** The mode of action of berninamycin and the mechanism of resistance in the producing organism, *Streptomyces bernensis, J. Gen. Microbiol.*, 128, 875, 1982.
147. **Teraoka, H. and Tanaka, K.,** Properties of ribosomes from *Streptomyces erythreus* and *Streptomyces griseus, J. Bacteriol.*, 120, 316, 1974.
148. **Weisblum, B.,** Induced MLS resistance: the molecular mechanism and its relation to antibiotic production, in *Proc. 4th Int. Symp. Genet. Ind. Microorganisms 1982,* Ikeda, Y. and Beppu, T., Eds., Kodansha, Tokyo, 1983, 33.
149. **Thompson, C. J., Ward, J. M., and Hopwood, D. A.,** DNA cloning in *Streptomyces:* resistance genes from antibiotic producing species, *Nature (London),* 286, 525, 1980.
150. **Thompson, C. J., Skinner, R. H., Thompson, J., Ward, J. M., Hopwood, D. A., and Cundliffe, E.,** Biochemical characterization of resistance determinants cloned from antibiotic-producing *Streptomycetes, J. Bacteriol.*, 151, 678, 1982.
151. **Thompson, C. J., Kieser, T., Ward, J. M., and Hopwood, D. A.,** Physical analysis of antibiotic-resistance genes from *Streptomyces* and their use in vector construction, *Gene,* 20, 51, 1982.
152. **Gale, E. F., Reynolds, P. E., Cundliffe, E., Richmond, M. H., and Waring, M. J.,** *The Molecular Basis of Antibiotic Action,* (Russian transl.), Mir, Moscow, 1975, 350.
153. **Mikulík, K., Karnetová, J., Křemen, A., Tax, J., and Vaněk, Z.,** Protein synthesis and production of tetracycline in *Streptomyces aureofaciens,* in *Radiation and Radioisotopes for Industrial Microorganisms,* International Atomic Energy Agency, Vienna, 1971, 201.
154. **Ludvík, J., Mikulík, K., and Vaněk, Z.,** Fine structure of *Streptomyces aureofaciens* producing tetracycline, *Folia Microbiol.*, 16, 479, 1971.
155. **Borisoglebskaya, A. N., Perebityuk, A. N., and Boronin, A. M.,** Study on resistance of *Actinomyces rimosus* to oxytetracycline (in Russian), *Antibiotiki,* 24, 883, 1979.
156. **Chopra, I. and Howe, T. G. B.,** Bacterial resistance to the tetracyclines, *Microbiol. Rev.*, 42, 707, 1978.
157. **Levy, S. B. and McMurry, L.,** Plasmid-determined tetracycline resistance involves new transport systems for tetracycline, *Nature (London),* 275, 90, 1978.
158. **Shales, S. W., Chopra, I., and Ball, P. R.,** Evidence for more than one mechanism of plasmid-determined tetracycline resistance in *Escherichia coli, J. Gen. Microbiol.*, 121, 221, 1980.

159. **Gale, E. F., Reynolds, P. E., Cundliffe, E., Richmond, M. H., and Waring, M. J.,** *The Molecular Basis of Antibiotic Action,* (Russian transl.), Mir, Moscow, 1975, 249.
160. **Yoshida, T., Weissbach, H., and Katz, E.,** Inhibitory effect of actinomycin upon the producing organism, *Arch. Biochem. Biophys.,* 114, 252, 1966.
161. **Marshall, R., Redfield, B., Katz, E., and Weissbach, H.,** Changes in phenoxazinone synthetase activity during the growth cycle of *Streptomyces antibioticus, Arch. Biochem. Biophys.,* 123, 317, 1968.
162. **Ochi, K.,** Control of actinomycin biosynthetic pathway in and actinomycin resistance of *Streptomyces* spp., *J. Bacteriol.,* 150, 598, 1982.
163. **Hitchcock, M. J. M. and Katz, E.,** Actinomycin biosynthesis by protoplasts derived from *Streptomyces parvulus, Antimicrob. Agents Chemother.,* 13, 104, 1978.
164. **Jones, G. H.,** RNA synthesis in *Streptomyces antibioticus: in vitro* effects of actinomycin and transcriptional inhibitors from 48-h cells, *Biochemistry,* 15, 3331, 1976.
165. **Gale, E. F., Reynolds, P. E., Cundliffe, E., Richmond, M. H., and Waring, M. J.,** *The Molecular Basis of Antibiotic Action,* (Russian transl.), Mir, Moscow, 1975, 322.
166. **Davies, J. E. and Yagisawa, M.,** The aminocyclitol glycosides (aminoglycosides), in *Biochemistry and Genetic Regulation of Commercially Important Antibiotics,* Vining, L. C., Ed., Addison-Wesley, Reading, Mass., 1983, 329.
167. **Cella, R. and Vining, L. C.,** Action of streptomycin on the growth of *Streptomyces griseus, Can. J. Microbiol.,* 20, 1591, 1974.
168. **Valu, G. and Szabó, G.,** Streptomycin sensitivity of ribosomes isolated from a streptomycin-producing *Streptomyces griseus, Acta Microbiol. Acad. Sci. Hung.,* 26, 207, 1979.
169. **Sugiyama, M., Kobayashi, H., Nimi, O., and Nomi, R.,** Susceptibility of protein synthesis to streptomycin in streptomycin-producing *Streptomyces griseus, FEBS Lett.,* 110, 250, 1980.
170. **Sugiyama, M., Mochizuki, H., Nimi, O., and Nomi, R.,** Roles of streptomycin 6-kinase and ribosomal affinity to streptomycin in self-protection of streptomycin producer, *Biotechnol. Lett.,* 3, 357, 1981.
171. **Sugiyama, M., Mochizuki, H., Nimi, O., and Nomi, R.,** Mechanism of protection of protein synthesis against streptomycin inhibition in a producing strain, *J. Antibiot.,* 34, 1183, 1981.
172. **Nimi, O., Sugiyama, M., Kameoka, H., Tomoeda, H., Ono, K., and Nomi, R.,** Fate of streptomycin in mycelium of the producer organism, *Biotechnol. Lett.,* 3, 239, 1981.
173. **Sugiyama, M., Mochizuki, M., Nimi, O., and Nomi, R.,** Assessment of competitive action of streptomycin 6-kinase and streptomycin 6-phosphatase in the *in vitro* protein synthesis of a streptomycin producing microorganism, *FEBS Lett.,* 139, 331, 1982.
174. **Sugiyama, Sakamoto, M., Mochizuki, H., Nimi, O., and Nomi, R.,** Purification and characterization of streptomycin 6-kinase, an enzyme implicated in self-protection of a streptomycin-producing microorganism, *J. Gen. Microbiol.,* 129, 1683, 1983.
175. **Hotta, K., Yamamoto, H., Okami, Y., and Umezawa, H.,** Resistance mechanisms of kanamycin-, neomycin-, and streptomycin-producing streptomycetes to aminoglycoside antibiotics, *J. Antibiot.,* 34, 1175, 1981.
176. **Piwowarski, J. M. and Shaw, P. D.,** Streptomycin resistance in a streptomycin-producing microorganism, *Antimicrob. Agents Chemother.,* 16, 176, 1979.
177. **Ono, H., Crameri, R., Hintermann, G., and Hütter, R.,** Hydroxystreptomycin production and resistance in *Streptomyces glaucescens, J. Gen. Microbiol.,* 129, 529, 1983.
178. **Crameri, R., Kieser, T., Ono, H., Sanchez, J., and Hütter, R.,** Chromosomal instability in *Streptomyces glaucescens.* Mapping of streptomycin-sensitive mutants, *J. Gen. Microbiol.,* 129, 519, 1983.
179. **Ogawa, Y. and Perlman, D.,** The occurrence of streptomycin-phosphorylating enzymes in streptomycin-producing streptomycetes, *J. Antibiot.,* 29, 1112, 1976.
180. **Yamamoto, H., Hotta, K., Okami, Y., and Umezawa, H.,** Self-resistance of a *Streptomyces* which produces istamycins, *J. Antibiot.,* 34, 824, 1981.
181. **Yamamoto, H., Hotta, K., Okami, Y., and Umezawa, H.,** Mechanism of resistance to aminoglycoside antibiotics in nebramycin-producing *Streptomyces tenebrarius, J. Antibiot.,* 35, 1020, 1982.
182. **Hotta, K., Yamamoto, H., and Okami, Y.,** Aminoglycoside antibiotic productivity of actinomycetes with reference to their multiple resistance to aminoglycoside antibiotics, in *Proc. 4th Symp. Genet. Ind. Microorganisms 1982,* Ikeda, Y. and Beppu, T., Eds., Kodansha, Tokyo, 1983, 97.
183. **Sugiyama, M., Nimi, O., and Nomi, R.,** Susceptibility of protein synthesis to neomycin in neomycin-producing *Streptomyces fradiae, J. Gen. Microbiol.,* 121, 477, 1980.
184. **Davies, J., Houk, C., Yagisawa, M., and White, T. J.,** Occurrence and function of aminoglycoside-modifying enzymes, in *Genetics of Industrial Microorganisms,* Sebek, O. K. and Laskin, A. I., Eds., American Society for Microbiology, Washington, D.C., 1979, 166.
185. **Hara, O. and Beppu, T.,** Induction of streptomycin-inactivating enzyme by A-factor in *Streptomyces griseus, J. Antibiot.,* 35, 1208, 1982.

186. **Okanishi, M. and Hotta, K.,** Regulation of antibiotics production in microorganisms, in *Molecular Breeding and Genetics of Applied Microorganisms,* Sakaguchi, K. and Okanishi, M., Eds., Kodansha, Tokyo, 1980, 9.
187. **Richmond, M.,** Antibiotic resistance and the evolution of bacteria, *Nature (London),* 302, 657, 1983.
188. **Walker, M. S. and Walker, J. B.,** Streptomycin biosynthesis and metabolism. Enzymatic phosphorylation of dihydrostreptobiosamine moieties of dihydrostreptomycin-(*streptidino*)phosphate and dihydrostreptomycin by *Streptomyces* extracts, *J. Biol. Chem.,* 245, 6683, 1970.
189. **Walker, J. B. and Skorvaga, M.,** Phosphorylation of streptomycin and dihydrostreptomycin by *Streptomyces, J. Biol. Chem.,* 248, 2435, 1973.
190. **Benveniste, R. and Davies, J.,** Aminoglycoside antibiotic-inactivating enzymes in actinomycetes similar to those present in clinical isolates of antibiotic-resistant bacteria, *Proc. Natl. Acad. Sci. U.S.A.,* 70, 2276, 1973.
191. **Davies, J. E.,** Plasmids in antibiotic-producing organisms: their roles in biosynthesis and resistance, in *Proc. 4th Int. Symp. Genet. Ind. Microorganisms 1982,* Ikeda, Y. and Beppu, T., Eds., Kodansha Ltd., Tokyo, 1983, 103.
192. **Koch, A. L.,** Evolution of antibiotic resistance gene function, *Microbiol. Rev.,* 45, 355, 1981.

Chapter 2

NUCLEIC ACID METABOLISM IN RELATION TO GROWTH

T. Ramakrishnan

TABLE OF CONTENTS

I. INTRODUCTION

For the purpose of this review, I shall be concentrating my attention on the growth of microorganisms. I shall define growth as a manifestation of the ability of the microbial cell to make protein, since protein makes up the bulk of the dry weight of the cell.

Schaechter et al.[1] were probably the first to report that in enteric bacteria, principally *Salmonella typhimurium,* the cellular mass and amounts of deoxyribonucleic acid (DNA), ribonucleic acid (RNA), and protein increase exponentially with the bacterial growth rate. *S. typhimurium* is able to synthesize all cellular components from a single carbon source and inorganic salts, and can utilize a number of organic compounds as carbon sources. Certain carbon sources such as acetate are metabolized slowly and support growth at a low rate; others, such as glucose, are metabolized more rapidly and support faster growth. If precursors of macromolecules, e.g., amino acids, are added to the medium, growth is even more rapid. Thus, by changing the composition of the medium, one can obtain growth of *S. typhimurium* with doubling times that vary between 20 min and 5 hr. With increase in the rate of growth, the cells become larger, contain more DNA, and often have an increased number of chromosomes. At a constant temperature, the increase in the amount of protein keeps pace with that of DNA, while the increase of cell mass and RNA is more rapid. In other words, the ratio of protein to DNA is independent of the growth rate, while the ratios of RNA to DNA and RNA to protein increase exponentially with the growth rate. This is understandable, since a cell growing at a fast rate must synthesize protein much more rapidly than does a slowly growing cell. This higher rate of protein synthesis requires that the cell contain more ribosomes, since the rate of protein synthesis per ribosome is constant.

Cooper and Helmstetter[2] have described the amount of DNA per cell at various growth rates in a model. According to this model, G(γ), the amount of DNA per cell at various growth rates, can be obtained from the following equation:

$$G(\gamma) = \frac{\gamma}{C\ln 2}\, 2^{(C+D)/\gamma} - 2^{D/\gamma}$$

where G = genome equivalents of DNA, γ = the doubling time, C = the time for replication of the bacterial chromosome, and D = the time between completion of a round of replication and cell division. From this equation it follows that the amount of DNA per cell increases in a nearly exponential manner with the growth rate. The equation assumes that the average rate of DNA replication is constant at any particular growth rate and that the C period is uninterrupted.

Donachie[3] has combined the observations of Schaechter et al. about the relationship of cell mass and growth rate with the model of Cooper and Helmstetter, and has shown that the cell mass per chromosome at the time of initiation of chromosome replication is nearly independent of the growth rate. This suggests that the mechanism regulating DNA replication is directly related to the cell mass.

Using *Escherichia coli* B/r, Dennis and Bremer[4] have re-examined the growth dependency of the bacterial RNA and protein content and the invariance of the protein to DNA ratio. They obtained low growth rates by employing a growth medium of poor nutritional quality and found that the accumulation of protein, rather than mass, is related to DNA replication. This implies that the exponential increase of cell mass with growth rate cannot be explained by the mechanism suggested by Donachie. The results of these authors suggest that the mass/cell curve represents the sum of the major molecular cell components (weights of protein + RNA + DNA). They also found that the ratios of RNA to DNA, RNA to protein, and protein to DNA give linear relationships with the growth rate (above 1.2 doublings per hour), whereas RNA to cell and protein to cell show a more complex growth rate dependency.

The synthesis of nucleic acids and protein in bacteria growing exponentially has been studied after a nutritional shift-up by Brunschede et al.[6] They studied the synthesis of these macromolecules in *E. coli* B/r for 2 hr after a nutritional shift-up from succinate minimal medium (growth rate $\mu_1 = 0.67$ doublings per hour) to glucose plus amino acids medium ($\mu_2 = 3.14$ doublings per hour). The long duration of 2 hr was necessary because a theoretical analysis carried out earlier[5] indicated this requirement for the transition between two steady states of balanced growth. During post-shift growth of the culture, the amounts of RNA (R), DNA (D), and protein (P) increased such that the ratios of the increments (R/P; D/P) were constants (k_1, k_2). This implies that the rates of accumulation of nucleic acids and proteins from time-invariant ratios were $(dR/dt){:}(dD/dt){:}(dP/dt) = k_1{:}k_2{:}1$. These constants change from their preshift value to their final post-shift value (i.e., k_1 and k_2) within a few minutes after the shift. k_1 is a function of the activity of ribosomes, whereas k_2 is related to the initiation of rounds of DNA replication. These parameters and the observed change in the doubling time of RNA ($= \mu_2/\mu_1$) were used to derive kinetic equations that describe the synthesis of DNA, RNA, and protein during the 2- to 3-hr transition period after a shift-up. (For details, please see the original paper.) The calculated kinetics agree closely with the observed kinetics.

The rate of synthesis of DNA and RNA has been studied in the slow-growing bacterium *Mycobacterium tuberculosis* by Ramakrishnan et al. and compared with that of the faster-growing *M. smegmatis* and of still faster-growing *E. coli*. Details of this work will be given later in the text under the appropriate heading.

In the following pages, I shall try to describe the role of each type of nucleic acid (and, in some cases, nucleic acid-metabolizing enzyme) in the growth of microorganisms, and also the regulation of growth by these nucleic acids.

II. DNA

Gudas and Pardee[7] measured the rates of DNA synthesis during the division cycles of *E. coli* strains B/r, K-12 3000, 15T, and 15 under several conditions of slow growth. They first synchronized the cultures by sucrose density gradient centrifugation of exponentially growing cultures and isolation of the smallest cells for measuring the rates of DNA synthesis. Their aim was to resolve the conflict in literature regarding the timing of a gap in the DNA synthesis during the cell cycle of *E. coli*. It had been observed by Helmstetter[8] and Helmstetter and Cooper,[9] using the membrane elution technique, that the gap was during the last one third of the cell cycle of *E. coli* B/r. Lark and his group[10,11] found that in *E. coli* $15T^-$ and *E. coli* B/r, it was at or near the beginning of the cell cycle. In all the strains of *E. coli* studied by Gudas and Pardee through the measurement of ^{3}H-thymidine uptake, the gap was found late in the cell cycle, prior to division, thus confirming Helmstetter's results. Thus, appear to regulate DNA synthesis in a similar manner during slow growth.

The conclusions of Gudas and Pardee have not been universally accepted by other workers who used different strains of *E. coli*. Helmstetter and Purucci[12] observed that the gap was during the last one third of the cell cycle when they used *E. coli* B/r A, but at the beginning of the cell cycle when they used *E. coli* B/r K and B/r F. More recent results obtained by Koppes et al.,[13] however, do not support qualitatively different patterns for strains B/r A and B/r K.

An explanation[14] for the different results of different workers is that normal DNA synthesis patterns are often perturbed by techniques of cell synchronization, and that the pertrubed patterns are not representative of steady-state cultures. Kubitschek and Newman[15] have employed a method to measure DNA synthesis periods in slowly growing exponential-phase and other steady state cultures — a method that avoids possible perturbations due to cell synchronization or selection techniques. The method relies upon the lethal effects of decay

Table 1
DNA CHAIN ELONGATION RATES

	M.tuberculosis $H_{37}R_v$	*M. smegmatis* SN2	*E. coli* B	*E. coli* B/r
Molecular weight of the genome	2.6×10^{9}[a]	4.5×10^{9}[b]	2.8×10^{8}[c]	2.8×10^{8}[c]
Length of the genome in kilo base pairs	3,900	6,800	4,300	4,300
Nature of the origin and fork	Single bidirectional[d]	Single bidirectional[d]	Single bidirectional[d]	
C period (minutes)	620	105	55	40
Chain elongation rate (number of nucleotides per minute)	3,200	32,400	39,000	53,750

[a] Reference 19.
[b] Reference 20.
[c] Reference 21.
[d] Assumed from DNA increment kinetic studies.

of radioisotopes incorporated in DNA — a method that has been used extensively in classical studies in bacteriophage and in rapidly growing bacteria. Kubitschek and Newman have extended this method to slowly growing cultures to determine the fraction of the cells that incorporates ^{125}I, and, therefore, are undergoing chromosome replication. In their experiments, *E. coli* B/r cells were exposed briefly to the thymidine analog 5-^{125}I iodo-2′-deoxyuridine and then stored in liquid nitrogen to accumulate lethal lesions resulting from radioactive decay. Values of surviving cell fractions were used in conjunction with values of DNA per cell to calculate C and D, according to an equation derived by them. Their results showed that the duration of the period of DNA synthesis in slowly growing, steady-state cultures was the same for all three strains of *E. coli* B/r used by them. In addition, while DNA synthesis can occupy the entire cell cycle in rapidly growing cultures, in very slowly growing cultures DNA synthesis was confined approximately to the final third of the cycle for each of the three strains.

The findings of Kubitschek and Newman appear to support an earlier conclusion[16] that very slowly growing *E. coli* cultures have division cycles formally analogous to those for eukaryotic cells, with stages similar to G_1, S, and G_2 periods. A similar conclusion has been reach by Hiriyanna and Ramakrishnan[17] working with the slow-growing microorganism, *Mycobacterium tuberculosis* $H_{37}R_v$.

Hiriyanna and Ramakrishnan have also compared the DNA chain elongation rate of *Mycobacterium tuberculosis* with that of a more rapidly growing mycobacterium like *M. smegmatis* and of *E. coli* which grows even more rapidly than *M. smegmatis*. They used the DNA increment, along with the data evaluation procedure of Bremer and Churchward,[39] referred to later, to determine the C period of the organisms, all growing at 37°C in the medium of Youmans and Karlson.[18] Their results are summarized in Table 1. It can be seen from the table that the C period of *M. tuberculosis* is about 10 hr, compared to about $1^1/_2$ hr for *M. smegmatis* and less than 1 hr for *E. coli*. These values were checked with the late log-phase synchronization method of Cutter and Evans.[22] The value of C period, according to this method, was found to be 9 hr for *M. tuberculosis* and 80 min for *M. smegmatis*. These values are close to the values obtained by the DNA increment method. In the calculation of chain elongation rates using the known values for the size of the mycobacterial and *E.*

coli genome, the authors have implicitly assumed a single replication origin and bidirectional replication fork. The results show that the slow growth of mycobacteria can be partly attributed to a slow rate of DNA chain elongation.

The overall rate of bacterial chromosome replication is believed to be regulated by the frequency of initiation and not by the rate of DNA synthesis at each fork.[2,23] Although the rate of DNA synthesis appears to be independent of the rate of growth at these generation times, as the generation time increases beyond 50 min, the transit time for replicating forks also increases.[9,10,24] An increase in the transit time for chromosome replication could be accounted for by a decrease in the rate of polymerization and/or a decrease in the initiation frequency for Okazaki fragment.[25] Although it has been shown that a limited supply of certain DNA precursors decreases the rate of DNA synthesis,[26,27] it is still not clearly understood which of these two possibilities is actually affected.

A third possibility that the slower rates of chromosome replication are due to a switch in the topology of replication from bidirectional to unidirectional has been investigated by Rodriguez and Davern.[28] They employed steady state ^{3}H-thymine incorporation studies along with DNA autoradiography to test this hypothesis in slow-growing cells of *E. coli*. In these experiments, slow rates of growth were achieved either by substituting for glucose the low energy-yielding carbon sources succinate and aspartate, or by the use of chloramphenicol at concentrations at which the rate of DNA synthesis is approximately one half that in untreated cells.[29] The results showed that the bidirectional mode of replication persists even under these conditions, and hence the slower rates of replication are not due to a switch in the topology of replication.

The time required for a DNA replication fork to traverse the chromosome from origin to terminus — the C period — is a fundamental parameter of bacterial growth. In exponentially growing cultures, together with the doubling time, it determines the number of copies of a gene relative to all genes and, since the expression of multiple copies of a gene is regulated, it also determines in part the activity of individual genes.[30] Further, it partly determines the timing of initiation and termination of rounds of replication relative to cell division and thus the size and macromolecular content of the average cell in a bacterial culture.[9]

Pritchard and co-workers[31,32] have devised two methods to study the regulation of DNA replication and to estimate the C period: the "step-up" method and the DNA increment method. The step-up method is based on the fact that in thymine-requiring bacteria the velocity of the chromosome replication fork can be varied by changing the thymine concentration of the medium. The authors made use of this phenomenon to carry out a step-up experiment in which they transferred bacteria from low to high thymine concentrations. In such an experiment it was found that accumulation of DNA immediately accelerated, although the mass doubling time was not affected. Later after the step-up, DNA accumulation returned to the doubling time of the mass, but the DNA mass ratio remained at the new steady state. The step-up is then distinct from, and complementary to, the nutritional "shift-up".[33] In the nutritional shift-up, the mass doubling time (τ) decreases,but the time required for chromosome replication (C) remains constant. In the Pritchard and Zaritsky step-up experiment, the C period decreases (since the replication fork velocity is increased in the higher thymine concentration), but τ is constant. Just as analysis of the nutritional shift-up has proven to be important in the control of RNA and protein synthesis,[34] analysis of the step-up experiment is important in studying the regulation of DNA replicaiton. In addition, since the step-up results in an altered DNA to mass ratio, but the net synthesis of RNA and protein is unaffected, there must be changes in the rate of transcription per DNA or per average gene. Thus, an understanding of the step-up is also important for the problem of gene transcription.

In the DNA increment method, Pritchard and Zaritsky[39] estimated the C period of *E. coli* $15T^{-}$ and CR 34 by measurement of the increase in DNA after cessation of initiation brought about by amino acid starvation. The value thus obtained for both strains was 57 min at "infinite" thymine concentration, which is much longer than that reported by Helmstetter and Cooper[2] for other strains.

Bremer et al.[34] have extended Pritchard and Zaritsky's analysis of the step-up experiment and have derived the theoretical relation between replication initiation, termination, and DNA accumulation for experiments in which the length of time required for replication of the bacterial chromosome (C period) can be varied. This theory enables one to determine absolute values of the C period from kinetics of DNA accumulation after a step-up experiment. Application of this method of data evaluation to an observed step-up experiment with a thy^- derivative of *E. coli* B/r indicated that the theory describes the observed post-step accumulation within experimental errors. It was found that changes in the replication velocity (C) do not measurably affect the timing of initiation events in culture.

Bremer and Churchward[35] have shown how the values of C obtained by Pritchard and Zaritsky for *E. coli* $15T^-$ by the DNA increment method can be brought closer to other strains if one takes into account a delay in the cessation of replication initiation after inhibition of protein synthesis. They have derived a theory describing the effect of inhibition of initiation rounds of DNA replication on the accumulation of DNA and have presented an analysis which allows the determiniation of the time C taken to replicate the bacterial chromosome from the kinetic changes in the accumulation of DNA. This analysis was applied to experiments in which inhibition of initiation was achieved by inhibiting protein, or RNA and protein, synthesis by chloramphenicol or rifampin. The results for both antibiotics are identical and indicate that there is a delay of 6 to 11 min in the effect of the antibiotics on initiation of rounds of replication. If this delay is taken into account, then the value of the C period estimated from such experiments agrees with values obtained by other methods, whereas by conventional data evaluation of such experiments, e.g., those of Pritchard and Zaritsky, the C period could be over-estimated. In the low thymine-requiring derivative of *E. coli* B/r ATCC 12407 used by Bremer and Churchward, the C period was found to be between 38 and 41 min for cultures growing with a mass doubling time of 29 min in glucose-amino acids medium, supplemented with 20 μg of thymine per milliliter.

The analysis described above to obtain absolute values of C with the step-up method has been used by Churchward and Bremer[36] to resolve the disagreement as to the variation in C with the growth rate: whereas experiments with age-fractionated *E. coli* B/r indicate that C decreases with growth rate up to one doubling per hour and then becomes constant;[9] gene frequency measurements with exponential cultures of *E. coli* K12 suggest that C remains constant at all growth rates.[37] Churchward and Bremer have measured C under a variety of growth conditions with both *E. coli* B/r and a low thymine-requiring derivative of that strain. In the thy^- strain, C was measured as a function of growth rate and exogenous thymine concentration either from step-up or chloramphenicol experiments. In the thy^+ parental strain, C was measured only as a function of the growth rate and only by the chloramphenicol method. The results indicated that C decreases continuously with the growth rate such that it approaches a constant value of approximately 37 min at high growth rates.

It is clear from the above that the initiation of chromosome replication is an important factor in determining the amount of DNA per cell in bacteria. Churchward et al.[38] have reported that the initation mass, defined as cell mass per origin of DNA replication (OD units at 460 nm of culture ÷ origins per milliliter of culture), reflects the intracellular concentration or activity of a hypothetical factor that controls initiation of chromosome replication in bacteria. In *E. coli* B/r, the initiation mass was found to increase twofold with increasing growth rate between 0.6 and 1.6 doublings per hour; at higher growth rates (measured up to 2.4 doublings per hour), it remains essentially constant.

The initiation mass is one of the factors that determines the cell size and DNA concentration. The DNA concentration, expressed as genome equivalents of DNA per unit mass (G/M), is given by Pritchard and Zaritsky[39] in the following equation:

$$G/M = \frac{1 - 2^{-C/\tau}}{M_0(C/\tau)\ln 2}$$

where M_0 is initiation mass, τ is mass doubling time; and C is reciprocal of replication velocity.

The studies on the regulation of macromolecular synthesis in eukaryotes carried out on the lines described above for bacteria have been reported in *Saccharomyces cerevisiae*,[40,41] *Tetrahymena pyriformis*[42] *Euglena gracilis*,[43] and *Prototheca zopfii*. They have confirmed in a general way the dependency, here also, of the relative levels of DNA, RNA, and protein on growth rate. Therefore, both prokaryotes and eukaryotes may exist in one of a continuum of steady states which are dictated by the growth rate. In *Prototheca zopfii*, which grows by multiple fission, the daughter cell number also depends on growth rate, and hence on the relative levels of DNA, RNA, and protein.

III. RNA

The transcription of DNA to RNA in bacteria is not only regulated by repressor-mediated control mechanisms and attenuation, but also by the rate of growth of the organism. The growth rate-dependent variation in the expression of genes through the regulation of transcription initiation frequency has been termed "metabolic regulation".[45] There has been a controversy for quite some time over whether the rate of transcription initiation in bacteria is limited by the concentration of DNA, as claimed by Koch,[46] or, alternatively, by the functional RNA polymerase.[47] Churchward et al.,[48] in an elegantly executed set of experiments, have concluded that transcription is limited by the concentration of functional RNA polymerase in the cytoplasm. These authors used *E. coli* B/r strain TJ K16, which has a lower DNA concentration due to a mutationally altered control of initiation of replication and is also a thymine auxotroph. The DNA concentration was varied by changing the replication velocity or by changing replication initiation in this mutant. Their results demonstrated that changes in DNA concentration (per mass) have no effect on the cell growth rate and the rates of synthesis (per mass) of stable RNA (rRNA, tRNA), bulk mRNA, or protein, or on the concentration of RNA polymerase (total RNA polymerase per mass). Changing the DNA concentration does, however, fully induce *lac* gene activity, used by the authors as a model for constitutive gene expression.

Manor et al.[49] were probably the first investigators to measure in vivo RNA chain growth rates in bacteria. They allowed *E. coli* to incorporate a radioactively labeled nucleotide precursor into its RNA chains for a certain period, and ascertained the average chain length during the labeling period by measuring the ratio of the total radioactivity incorporated into the RNA to that present in the 3′ ends of the growing chains. Such measurements provide a means of calculating the nucleotide step time, which is the time elapsed between the addition of successive nucleotide residues to the growing 3′ ends of the RNA. An overall rate of RNA chain growth can be estimated by using an averaged value of the individual step times. The authors compared the average RNA chain growth rates of *E. coli* growing at 29°C in a glucose-casamino acids medium with a generation period of 1.07 generations per hour with that growing at 37.5°C. At 29°C, the RNA chain growth was estimated to be 26 nucleotides per second. At 37.5°C, both the generation period and nucleotide step times were found to be 60% of the corresponding values at 29°C.

The nucleotide step-times of *E. coli* growing at 29°C in a succinate medium with 0.63 cell generations per/hour and in a proline medium with 0.33 cell generations per/hour were found to be longer than the corresponding step times during the faster growth in glucose-casamino acids medium. The increases in nucleic acid step times with increases in generation period observed by the authors were not nearly great enough, however, to account for the great variations in the rate of RNA synthesis observed under the three different physiological conditions. The authors conclude that bacteria appear to adjust their specific content of RNA according to their physiological needs by varying both the rate of RNA chain growth and the number of nascent RNA molecules under synthesis at any time.

In a more direct approach, Dougan and Glaser[50] estimated the rate of chain elongation of ribosomal RNA molecules in *E. coli* by measuring the incorporation of radioactive guanine to rRNA and from these data estimated the transit time to transcribe 16 and 23S RNA. The times required to produce the individual molecules were determined by analysis of the labeling kinetics. The experiments were carried out under two different growth conditions using glucose and succinate as carbon sources, respectively. They found that rRNA transcription proceeds at about 84 nucleotides per second and with glucose and about 48 nucleotides per second with succinate. They also found that the total magnitude of rRNA synthesis in the cell depends on the carbon source, as does the transcription rate. However, the transcription rate does not vary sufficiently to account for the different rates of rRNA synthesis. The authors also conclude that the overall magnitude must, therefore, be controlled by the rate of chain initiation.

Bremer et al.[51] have made use of the drug rifampin, which inhibits RNA chain initiation, but not elongation, to estimate the rate of RNA chain growth in *E. coli*. This method allows one to arrive at the growth rate of the stable as well as the unstable RNAs. Using this method, Dennis and Bremer[52] measured the chain growth rate for ribosomal RNA (rRNA) growing in succinate (μ = 1.36 doublings per hour), and glucose plus amino acids (μ = 2.10 doublings per hour) media at 37.5°C. With increasing bacterial growth rate, the RNA chain growth rate was found to increase from about 75 to 100 n/min. These values are two- to threefold higher than the chain growth rates reported by Manor et al.[49] for total RNA and probably explains the inability of the latter to correlate the increase in total RNA to increase in growth of the bacteria. Dennis and Bremer's data imply that, following a nutritional shift-up, the transfer of a relatively small number of RNA polymerase molecules from unstable to stable RNA genes, along with the increase in the stable RNA chain growth rate, is sufficient to account for the abrupt increase in the net rate of RNA synthesis. In addition, their calculations indicate that the linear density of polymerase molecules on their DNA template increases with the bacterial growth rate such that in the growing bacteria all rRNA genes (48 copies at μ = 3.0 doublings per hour) are nearly saturated with RNA polymerase.

From the above data, Bremer et al.[53] calculated the fraction of RNA polymerase engaged in the synthesis of stable RNA, Ψ S, to be 0.22, 0.36, and 0.48 for growth in succinate, glucose, and glucose plus amino acid media, respectively, taking into account the difference between RNA chain growth rate of stable and unstable RNA. The authors conclude that (1) the increase in the net rate of RNA synthesis after a shift-up results from a transfer of RNA polymerase molecules from unstable to stable RNA genes and a concomitant increase in the stable RNA chain growth rate, but does not require an activation of RNA polymerase, and (2) the synthesis of functional RNA polymerase enzymes is subject to a growth rate-dependent control. These conclusions have been confirmed by Dennis and Bremer[54] by their finding that following a nutritional shift-up, both the fraction of functional RNA polymerase engaged in the synthesis of stable RNA, Ψ S, and the rRNA chain growth rate, C_s, increase within 5 min to near their final post-shift steady state values. The increase in these two parameters is sufficient to account completely for the observed sudden increase in the rate of RNA accumulation.

The increase in the bacterial growth as a result of temperature shift-up does not appear to result in an increase in the net rate of RNA synthesis, unlike in the case of the nutritional shift-up descibed above. Ryals et al.[55] found that for *E. coli* B/r growing in glucose minimal medium, the growth rate increased 3.4 fold between 20 and 40° C, but rRNA and tRNA concentration (RNA/mass), RNA polymerase concentration (RNA polymerase/mass), RNA polymerase activity (fraction of RNA polymerase activity engaged in RNA chain elongation), and stable RNA synthesis relative to total RNA synthesis remained invariant. There was, however, a 3.4 fold increase of guanosine tetraphosphate (ppGpp) concentration and rRNA chain elongation rate. ppGpp is well known to regulate RNA synthesis in bacteria, [56] and its apparent lack of effect on RNA synthesis under temperature shift conditions when its concentration is increased, observed earlier by Gallant et al.,[57] led the latter to believe that ppGpp is not involved in the control of stable RNA synthesis. However, Ryals et al. have concluded from their experiments that the mechanism of RNA synthesis by ppGpp, assumed to involve an interaction of RNA polymerase with ppGpp,[58] is itself temperature dependent such that, with increasing temperature, higher concentrations of ppGpp are required to affect the RNA polymerase.

Instead of changing the growth rate by either a nutritional or a temperature shift-up and investigating the effects of the changed growth rate on the rate of RNA chain growth, Harshey and Ramakrishnan[59] have compared the rate of RNA chain growth in a slow growing organism like *Mycobacterium tuberculosis* with that in *E. coli*, employing the methods of Manor et al. [49] and also the method of Bremer et al.[51] They found that the average rate of RNA chain elongation in the former was 4 nucleotides per second, which is about 10 times lower than that in *E. coli*. The ribosomal RNA chain growth rate was found to be 10 nucleotides per second, which, again is about 10 times lower than the corresponding rate in *E. coli*. Examination of the RNA content per DNA in *E. coli* and *M. tuberculosis* has shown that the latter contains a tenfold lower amount of RNA per genome (assuming that the size of the genome is the same in both organisms). Since *M. tuberculosis* undergoes 20-fold fewer divisions per hour than *E. coli*, it follows that the net rate of RNA synthesis per DNA template is $20 \times 10 = 200$-fold lower in this organism than in *E. coli*. Thus, a 10-fold decrease in the RNA chain elongation rate does not wholly explain the rather large difference in the net RNA synthesis rates between these organisms. (On the other hand, the lower DNA chain elongation rate in *M. tuberculosis*[32] referred to earlier in this chapter could partially contribute to this large difference.) It would appear, therefore, that such marked changes in the rate of RNA synthesis must reflect the initiation frequency of RNA molecules, as well as the number of RNA polymerase molecules actively engaged in RNA synthesis at a specified time.

The net rate of RNA synthesis is a function of both the synthesis and breakdown of RNA, which occur simultaneously. Burton[60] has analyzed the RNA turnover in *E. coli* using histidine as a radioactivity trap for [2-^{3}H] adenine nucleotides. ^{3}H was incorporated from [2-^{3}H] adenine into histidine and then into protein by bacterial cultures synthesizing histidine. The kinetics of incorporation of ^{3}H into protein, RNA, and DNA were used to evaluate RNA turnover in *E. coli*. In *E. coli* B at 37°C, the fraction of pulse labeled RNA, which is unstable, changes from 0.83 to 0.50 as the growth rate is increased by nutritional shift-up from 0.08 to 2.6 generations per hour. The unstable fraction is 0.65 at 1.27 generations per hour in glucose-salts medium. The author's results for other strains of *E. coli*, *Salmonella typhimurium*, and *Bacillus subtilis* indicate a similar extent of turnover.

Although RNA in mature ribosomes is not degarded in growing *E. coli*,[61] it is conceivable that surplus rRNA is synthesized and that the excess is degraded soon after synthesis. This is consistent with the extensive incorporation of [^{18}O] into phosphate groups of RNA during growth in [^{18}O] water [62] and with independent arguments that degradation of rRNA and tRNA might be important at least in slow growing *E. coli*.[63]

The turnover of RNA has also been investigated in the photosynthetic bacterium, *Rhodopseudomonas spheroides*,[64,65] an organism which can grow well under both aerobic and anaerobic conditions. The authors have studied differences in the rate and extent of degradation of RNA labeled by a 30-sec pulse, under both conditions, using rifampin to block

RNA synthesis. In anaerobic cultures, unstable RNA is degraded with a half-life of 1.25 to 2.0 min, and about 40% of the pulse labeled RNA is stable. In aerobic cultures, the half-life of unstable RNA is increased to 2.5 to 4.0 min, and 50% of the RNA is stable. Going from aerobic to anaerobic environment, RNA begins to be broken down at a faster rate, which increases along with an increase in the extent of degradation for at least 60 min. In an anaerobic to aerobic transition, no increased RNA instability is noticed for 30 min, but by 90 min, both the extent and rate of breakdown are reduced. These alterations in stability are not reflected in the total RNA synthesis, since only a small fraction of the cellular RNA is involved. Through the use of puromycin and chloramphenicol the authors have shown that the change in RNA is probably connected with regulation of specific protein synthesis at the level of translation. They have also shown by the RNA-DNA hybridization method that identical species of RNA in the same relative amounts are synthesized by *R. spheroides* during aerobic and anaerobic growth on the same medium. In *E. coli* grown in a succinate medium, and therefore capable of only aerobic growth, transfer to anaerobic conditions leads to a protection of mRNA from degradation, presumably due to energy starvation.[66] It is obvious that this mechanism cannot be operative in *R. spheroides*, since during the aerobic to anaerobic transition, where a transient energy starvation does occur,[66] RNA becomes less stable. It can be concluded that in photosynthetic bacteria, light-specific protein synthesis (anaerobic conditions) is regulated at the level of translation, and this contributes to the difference in the half-life of unstable RNA under aerobic and anaerobic conditions of growth.

Net RNA synthesis is terminated during the growth restriction of *Yersinia pestis* in a calcium-deficient medium at 37°C.[67] This organism requires 2.5 m*M* Ca^{++} for growth at 37°C, but not at 26°C (Ca^{++} requirement is plasmid mediated and Ca^{++} is essential for expression of virulence by *Y. pestis*). After a shift from 26°C to 37°C in a Ca^{++}-deficient medium, an ordered series of metabolic alterations occur, which results in a transition from a growing cell to a viable but nonproliferating cell. The earliest known alteration in normal metabolism associated with this transition is a termination of net RNA synthesis. Competitive RNA-DNA hybridizations with uniformly labeled and stable RNA competitor indicated identical mRNA to stable RNA ratio in growing cells and nonproliferating Ca^{++}-deprived cells. Similar hybridizations with pulse labeled RNA demonstrated that growing cells synthesized 57% mRNA, 37% rRNA, and 5% tRNA, whereas Ca^{++}-deprived cells synthesized 95% mRNA, 4.7% rRNA, and 0.7% tRNA. After addition of radioactive uracil and rifampin to growing and Ca^{++}-deprived cells, decay of approximately 40 and 90% of the newly synthesized RNA was found for growing and Ca^{++}-deprived cells, respectively. The half-life of the mRNA was found to be 1.5 min for growing cells and 4.5 min for Ca^{++}-deprived cells. *Y. pestis* elicited increases of guanosine tetra- and pentaphosphates in response to amino acid deprivation and yielded transient increases in the levels of these phosphorylated nucleotides after a shift from 26 to 37°C. These increases were independent of Ca^{++} availability and preceded the alteration in RNA synthesis by more than 1 hour. The levels of these phosphorylated nucleotides than stabilized at about 80 and 40 pmol for Ca^{++}-deprived and Ca^{++}-supplemented cultures, respectively, and did not increase further in Ca^{++}-deprived culture at the time corresponding to the reduction in stable RNA synthesis. These findings indicate that the early lesion in RNA synthesis associated with the growth restriction of Ca^{++}-deprived *Y. pestis* reflects a block in stable RNA synthesis and that this effect is not mediated by guanosine tetra- or pentaphosphate.

IV. rRNA AND tRNA

One of the earliest observations [68,69] on the effects of growth of bacteria when shifted from a nutritionally poor to a rich medium is that this accelerated growth involves an increased synthesis of ribosomes. It is now known[4] that in this process there is a greater extent of tRNA charging with amino acids which reduces ribosome idling and thus synthesis of guanosine tetraphosphate.[70,71] A low concentration of this nucleotide is presumed to increse the expression of rRNA and ribosomal protein genes,[72-74] thereby stimulating ribosome synthesis.

The fraction of ribosomes found in polyribosomes is generally taken as representing the fraction engaged in protein synthesis. In *Escherichia coli*, the fraction of ribosomes in polyribosomes has been reported to increase with increasing specific growth rate.[75] Harvey[76] has estimated the fraction of ribosomes engaged in protein synthesis in *E. coli* growing at specific growth rates ranging from 0.3 to 1.2 hr^{-1} from measurements of nascent protein/ribosome and of polyribosomes, in nutritional shift-up experiments. Both measurements showed that the fraction of ribosomes synthesizing proteins increased with specific growth rate, from 30 to 70% over the range studied. Polyribosome measurements made at different specific growth rates in four *E. coli* strains showed no significant differences between strains. The increase in the fraction of polyribosomes had begun within 30 sec after a shift-up from glucose minimal medium, and was complete within 2 to 5 min. These results indicate that the function of ribosomes is regulated. The increase in the fraction of ribosomes synthesizing protein coincides with the increase in the rate of ribosomal protein synthesis during a time when the rate of synthesis of other proteins increases hardly at all.[77] This suggests that a major purpose of regulation of ribosomal function is control of protein synthesis.

The rate of ribosome production itself has been measured in *Bacillus licheniformis* subjected to nutritional shift-up.[78] The results indicate that the ribosome content per cell decreased considerably with increase in generation time. Nevertheless, the time required for the biosynthesis of a single 50S subunit was constant (about 10 min) and independent of the specific growth rate of the cell (within the generation time limits of 35 to 120 min). The authors conclude that these results show that ribosome production is not controlled by the rate of assemblage of individual ribosomes, but by the number of ribosomes manufactured at the time.

Almost similar conclusions have been drawn by Varrichio and Monier[79] working with *E. coli*. They found that there is no correlation between ribosome subunit levels and growth rate. Under equilibrium conditions and within the generation time limits of 0.5 and 1.35 μ, the percentage of subunits to total ribosomes is constant at about 20%. Furthermore, the polysome content, when related to DNA content, varies directly with the growth rate, which indicates that the average efficiency of polysomes in protein synthesis does not vary over the range of growth rates tested.

In contrast to the above findings, Michealis[80] has reported that in *E. coli* subjected to nutritional shift-up, the rate of ribosome assembly varies directly with the growth rate. It is possible that the different observations may stem from the differences in experimental design. The finding of Michaelis suggests that the number of ribosomes characteristic of a particular growth rate is regulated at least in part by the rate of ribosome assembly.

The synthesis of tRNA and rRNA appears to be regulated by similar yet different mechanisms.[81] The authors analyzed the cellular levels of tRNA as a function of the growth rate of *E.coli* $15T^-$ and also attempted to determine whether the synthesis of isoaccepting species of tRNA are regulated coordinately during fast and slow growth, using broth, glucose, succinate, and acetate media. With decreasing growth rates, the rRNA fraction of total RNA decreased slightly, but the tRNA fraction increased. The net result of these changes is that the cellular contents of both tRNA and rRNA decreased with decreasing growth rates. For several amino acid-specific tRNAs studied, acceptor capacities per unit DNA increased with increasing growth rate. Isoacceptors for individual amino acids appear to be regulated as a group as evidenced by similar isoaccepting profiles for a number of amino acids for rapidly (broth) and slowly growing (acetate culture) cells. The authors also found that the tRNA efficiencies in protein synthesis showed an increase of about 200% between 0.5 and 1.7

generations per hour, whereas the corresponding rRNA efficiencies were only about 100%. They therefore conclude that tRNA appears to be a more reasonable limiting agent in protein synthesis than ribosomes. Also, since the tRNA to DNA ratio both for bulk tRNA and specific tRNAs changes with growth rate, the amount of tRNA per cell must be determined by a regulatory mechanism and not by gene dosage.[82]

Total RNA content also decreases with decreasing growth rates in a number of eukaryotic microorganisms, namely, the yeast *Saccharomyces cerevisiae*,[83-85] the protozoan *Tetrahymena pyriformis*,[86] and the alga *Prototheca zoffii*.[87] The effect of growth rate on the relative amounts of rRNA and tRNA has been studied with *T. pyriformis* and *S. cerevisiae*. The results obtained are similar to those from bacteria: the two types of RNA decrease to slightly different exents with decreasing growth rate. The noncoordination of rRNA and tRNA levels may be due to controls operating on two distinct polymerases, one of which synthesizes rRNA and the other tRNA. The heterogeneity of RNA polymerase in bacteria has already been reported,[88,89] while the presence of three types of RNA polymerases in eukaryotic cells has been well documented.[90,91]

V. DNA-DEPENDENT RNA POLYMERASE

It has already been mentioned that Bremer et al.[53] concluded from their experiments on nutritional shift-up of *E. coli* that the synthesis of functioning RNA polymerase enzymes is subject to a growth rate-dependent control. Probably the first person to attempt to measure the synthesis of RNA polymerase in growing *E. coli* was Dalbow.[92] He based his experiments on the finding of Matzura et al.[93] that if a crude lysate of *E. coli* is subjected to SDS gel electrophoresis, the only two proteins discernible of high molecular weight (>140,000) are the β- and β′-subunits of RNA polymerase, and investigated the synthesis of these subunits as a function of bacterial growth rate. In succinate (μ = 0.67 doublings per hour), glucose (μ = 1.36 doublings per hour), and glucose plus amino acids (μ = 2.10 doublings per hour) supplemented media, the fractions of [^{14}C] leucine-labeled β- and β′- protein/total protein were found to be 1.05, 1.31, and 1.56%, respectively. He then compared these values with estimates from his laboratory of the differential rate of synthesis of functional RNA polymerase. The results suggested an excess of total over functional RNA polymerase. He concluded from these that the cell may synthesize more β- and β′- subunits than α- or σ- subunits and that the synthesis of RNA polymerase may not be completely coordinated. He concludes, therefore, that the amounts of β- and β′- subunits cannot be used to estimate the amount of RNA polymerase.

Dalbow's conclusions are at variance with recent findings[94,95] that there is a single promotor in *E. coli* for the genes of the subunits or RNA polymerase and that ribosomal protein genes are cotranscribed with RNA polymerase genes. These findings have suggested to many workers that RNA polymerase regulates the number of ribosomes after a nutritional shift-up.

Bremer[96] has estimated the parameters affecting the synthesis of ribosomes and RNA polymerase in bacteria. According to him, both syntheses are linked and the synthetic rates of both depend upon the values of six parameters: (1) the fraction of total RNA polymerase that is functional; (2) the fraction of total ribosomes that is functional; (3) the fraction of functional RNA polymerase that is engaged in rRNA synthesis; (4) the fraction of total protein that is RNA polymerase protein; (5) peptide chain elongation rate; (6) rRNA chain elongation rate. If these parameters are constant with time, then the numbers of both ribosomes and RNA polymerase molecules increase exponentially. Bremer has also shown how the rate constant (fractional increase per unit of time) relates to these parameters and how the kinetics of ribosome and RNA polymerase synthesis respond to change in any of these parameters.

In a series of elegant papers, Ishihama and co-workers[97-100] have investigated the biosynthesis of RNA polymerase in *E. coli* during growth. They also used the method of polyacrylamide gel electrophoresis of RNA polymerase in the presence of SDS and determined the intracellular amounts of the individual subunits by measuring the amounts of precipitates formed with specific antibody against the holoenzyme. However, they followed the biosynthesis of RNA polymerase by estimating only the β- subunit. This was because when they followed the procedure of Matzura et al.,[93] they found a polypeptide unrelated to the polymerase migrating in the immediate vicinity of the β'- subunit, interfering with the accurate measurement of this subunit. They also measured the rate of synthesis of individual subunits of RNA polymerase as a safety measure of the accuracy of their results. They found that when *E. coli* β/r grown in a succinate medium was subjected to shift-up by adding a mixture of glucose and amino acids, there was a rapid rise of the differential rates of the synthesis of α-, β-, and β'- subunits, leading to the increase of core polymerase concentration. The differential rates decreased thereafter to the level characteristic of the post-shift rate of cell growth. On the other hand, upon shift-down of the strain from a glucose-amino acids medium to a glucose medium lacking amino acids, the differential rate decreased rapidly and then regained the characacteristic of the new growth rate. (The rate of subunit synthesis, however, exhibited little change during shift-up, but decreased during shift-down.)

Ishihama and co-workers found a similar control on the rate of ribosomal protein synthesis, suggesting the coordinate expression of genes for the core polymerase subunits and ribosomal proteins. They conclude from their results that the intracellular concentration of RNA polymerase as well as of ribosomes might be one of the most important factors that affect the rate of bacterial growth.

Ishihama's group also determined the synthesis rate, content, and intracellular distribution of RNA polymerase for wild-type cells of *E. coli* under various phases of growth. The differential rate of synthesis decreases during the transition to the stationary phase, first for the α- subunit of RNA polymerase and ribosomal proteins, and subsequently for β- and β'-subunits, whereas the rate of σ- subunit synthesis remains virtually constant throughout the growth period. (There is a certain period of growth transition from exponential to stationary phase when twice as much β- subunit is synthesized as β'- subunit — the coordinated synthesis of RNA polymerase subunits appears to be temporarily broken — and the overproduced β subunit accumulates as the intermediate $\alpha_2\beta$- complex, which is rapidly and preferentially degraded. This repression of the synthesis of the core enzyme units can be mimicked by the depletion of glucose as a sole carbon source, and hence is brought about by the reduction in the concentration of glucose when the organism enters the stationary phase.) In consequence, the synthesis ratio among RNA polymerase subunits varies, depending on the phase of cell growth. In the exponentially growing phase, α- and σ- subunits are synthesized in excess of, or less than, β- and β'-subunits, respectively, on a molar basis. In contrast, nearly stoichiometric amounts are synthesized for all four subunits (1α, 1β, $1\beta'$, and 1σ) in the stationary phase. The decrease in the synthesis rate of core enzyme subunits as well as the degradation of both newly synthesized and pre-existing β- and β'-subunits leads to the reduction of the intracellular content of RNA polymerase core relative to the bulk of proteins as much as approximately 1.5-fold. In addition, the activity of RNA polymerase is repressed in the stationary phase, as indicated by the decrease of the nucleoid-bound RNA polymerase, and hence of the functioning polymerase, as much as 1.5-fold. The authors explain the decrease of the overall transcription activity by the reduction in both the concentration and activity of RNA polymerase. They also conclude from their results that the observed coupling between the repression of RNA polymerase synthesis and the increase of unused polymerase supports the autogenous regulation of RNA polymerase gene (rpo) expression.

The findings of Ishihama and co-workers that the fractional rate of RNA polymerase synthesis increases with growth rate as much as the fractional rate of ribosomal protein synthesis are at variance with the earlier reported observations by Engback et al.[101] that it is independent of the growth rate, and the report by Matzura et al.[102] that it increases with the growth rate but is less than the functional rate of ribosomal protein synthesis. The variations could be due to the fact that the various authors used different strains under different conditions. Shepherd et al.[103] remeasured the synthesis of RNA and of RNA polymerase simultaneously in the same cultures using an improved method of quantitative measurement of β- and β′- subunits in crude lysates of *E. coli* B/r (ATCC 12407). They also determined the concentration of total protein in the lysate and converted the ratio of β′- protein to total protein to fractional synthesis rate of RNA polymerase by multiplying with 2.4 [= molecular weight ratio of core RNA polymerase enzyme (398 kdaltons) to β′-subunit (165 kdaltons)].[104] This assumes that the amount of core polymerase is limited by the synthesis of β- and β′- subunits, since the cell contains an excess of the α- subunit.[105,106] The authors found that for a 3-fold increase in the growth rate of *E. coli*, the fractional synthesis of polymerase (relative to protein) increased 1.5-fold, whereas the fractional synthesis of ribosomal protein increased 2.2-fold. They interpreted the decrease in the amount of RNA polymerase with respect to ribosome with increasing growth rate — a confirmation of the findings of Matzura *et al.* referred to earlier — as an expression of the control of the transcriptional read-through from the genes for ribosomal proteins, rpl J, L, to the adjacent genes for RNA polymerase subunits, rpo B, C. They also determined the number of active RNA polymerase molecules from the synthesis rates of stable and messenger RNA and the known RNA chain growth rates. Comparison of active and total RNA polymerase indicated that the fraction of active enzyme increases from 20 to 30% in the range of growth rates between 0.6 and 2.0 doublings per hour. One of the possible causes for the inactive enzyme may be the fact that most of the core enzyme is bound to DNA rather than to σ factor.[107] Hence, changes in either the concentration of DNA or of σ factor could affect the RNA polymerase activity.

In a subsequent paper, Shepherd et al.[108] reported their results on the synthesis and function of RNA polymerase and ribosomes in *E. coli* B/r after a nutritional shift-up. The object of their experiments was to resolve the contradiction that, in spite of the fact that RNA polymerase genes are cotranscribed with ribosomal protein genes, there is a decrease in the amount of RNA polymerase with respect to ribosome with increasing growth rate, referred to earlier. They observed that two parameters affecting rRNA synthesis, i.e., the RNA polymerase activity and the relative rate of stable RNA synthesis (relative to total RNA synthesis), are actually not responding in a one-step manner, but are changing over a period of more that 1 hr after the shift-up from succinate-minimal to glucose-amino acids medium before they reach a final level characteristic for the postshift growth medium. This suggested that the distribution of transcribing RNA polymerase over ribosomal and nonribosomal genes and the polymerase activity are continuously adjusted to some growth-limiting reaction whose rate increases exponentially. The authors proposed that this reaction is the production of aminoacylated transfer RNA and that its exponentially increasing rate results in part from a gradually increasing concentration of aminoacyl transfer RNA synthetases after a shift-up. A computer simulation of a nutritional shift-up supported their idea.

RNA polymerase subunit biosynthesis under steady-state growth conditions has also been studied in *Bacillus subtilis*.[109] The authors determined the stoichiometries of α:β:β′:σ from cultures pulse-labeled during steady growth. The results suggested that in *B. subtilis* an unassembled pool of the α subunits exists from which the holoenzyme is formed. Upon shift-up from acetate to glycerol containing medium, a rapid rise in the differential rate of core enzyme synthesis was observed, while the rate of synthesis of the α- subunit was not stimulated. During shift-down from glycerol- to acetate-containing medium, a concomitant reduction in the synthesis of all subunits occurred for the first 20 min after the shift; thereafter, a rate of synthesis characteristic of the new growth rate was established. As cultures entered sporulation, an immediate reduction in the rate of β- and β′- subunit synthesis was found to take place.

VI. TRANSFER RNA METHYLTRANSFERASE

During the last few years the mechanism of the biosynthesis of tRNA has been intensively studied. The primary transcription product is cleaved by endo- and exonucleases and modified by methyltransferases. The growth rate-dependent regulation of tRNA methyltransferases has been studied by Ny and Bjork.[110] Their results indicate that tRNA (5-methyluridine) methyltransferase increases with growth rate in the same fashion as stable RNA. On the other hand, the activity of two other tRNA methyltransferases, i.e. 1-methylguanine methyltransferase and 5-methylaminomethyl-2-thiouridine methyltransferase, remains constant in relation to the growth rate. Thus, these three tRNA biosynthetic enzymes are not coordinately regulated. The reason for this could be that 5-methyluridine (ribothymidine) is present in all tRNA chains in *E. coli* and is always positioned in the TΨCG-loop of the tRNA,[111] while the other two are present only in 12 and 4% of the tRNA chains, respectively.[112] Regulation of tRNA (5-methyluridine) methyltransferase and stable RNA was found to be similar during shift-up and shift-down experiments. Thus, the regulation of tRNA (5-methyluridine) methyltransferase shares most of the hypothetical regulatory elements with those of the regulation of stable RNA and proteins involved in the translational machinery.

VII. CONCLUDING REMARKS

The information presented here shows that a significant amount of work on the relationship of nucleic acid metabolism to microbial growth has been carried out during the last 2 decades. The role of RNA in growth has been thoroughly worked out and the mechanism by which the regulation of growth by RNA is effected, viz., through the action of RNA polymerase and transfer RNA methyltransferase, has been well elucidated. Though similar investigations have been carried out on the role of DNA in growth, the enzymes regulating the action of DNA in growth have not yet been studied. However, the recent report[113] that the operon which encodes the sigma subunit of RNA polymerase also encodes ribosomal protein S21 and DNA primase in *E. coli* marks a beginning in the link-up of a DNA-metabolizing enzyme in growth. A systematic approach to the role of DNA-metabolizing enzymes in growth is needed. The development of a suitable method for estimating DNA polymerase III in crude bacterial lysates would greatly stimulate research in this important area. It is also highly desirable that work on these aspects is carried out by different workers on the same strain of bacteria, e.g. *E. coli* under identical conditions, so that conflicting results are not reported.

For a general review of the regulation of bacterial growth, RNA, and protein synthesis, the reader is referred to the article by Nierlich;[114] for a general review of bacterial growth and division, see the article of Mendelson.[115] A recent review by Rohl[116] gives a more detailed survey of the biosynthesis of bacterial ribosome and regulation.

ACKNOWLEDGMENT

The author is indebted to Dr. M. S. Shaila for her critical comments on the manuscript.

REFERENCES

1. **Schaechter, M., Maaloe, O., and Kjeldgaard, N. O.,** Dependence on medium and temperature of cell size and chemical composition during balanced growth of *Salmonella typhimurium, J. Gen. Microbiol.,* 19, 592, 1958.
2. **Cooper, S. and Helmstetter, C. E.,** Chromosome replication DNA synthesis and division cycle of *Escherichia coli* B/r, *J. Mol. Biol.,* 31, 519, 1968.
3. **Donachie, W. D.,** Relationship between cell size and time of initiation of DNA replication, *Nature (London),* 219, 1077, 1968.
4. **Dennis, P. P. and Bremer, H.,** Macromolecular composition during steady-state growth of *Escherichia coli* B/r, *J. Bacteriol.,* 119, 270, 1974.
5. **Bremer, H. and Dennis, P.,** Transitional period following a nutritional shift-up in the bacterium, *Escherichia coli* B/r: stable RNA and protein synthesis, *J. Theor. Biol.,* 53, 115, 1975.
6. **Brunschede, H., Dove, T. L., and Bremer, H.,** Establishment of exponential growth after a nutritional shift-up in *Escherichia coli* B/r: accumulation of deoxyribonucleic acid, ribonucleic acid and protein, *J. Bacteriol.,* 129, 1020, 1977.
7. **Gudas, L. J. and Pardee, A. B.,** Deoxyribonucleic acid synthesis during the division cycle of *Escherichia coli*: a comparison of strains B/r, K-12, 15 and $15T^-$ under conditions of slow growth, *J. Bacteriol.,* 117, 1216, 1974.
8. **Helmstetter, C. E.,** Rate of DNA synthesis during the division cycle of *Escerichia coli*, B/r, *J. Mol. Biol.,* 24, 417, 1967.
9. **Helmstetter, C. E. and Cooper, S.,** DNA synthesis during division cycle of rapidly growing *Escherichia coli, J. Mol. Biol.,* 31, 507, 1968.
10. **Lark, C.,** Regulation of DNA synthesis in *Escherichia coli*: dependence on growth rates, *Biochim. Biophys. Acta,* 119, 517, 1966.
11. **Chai, N. C. and Lark, K. G.,** Cytological studies of deoxyribonucleic acid replication in *Escherichia coli* $15T^-$: replication at slow growth rates and after a shift-up into rich medium, *J. Bacteriol.,* 104, 401, 1970.
12. **Helmstetter, C. E. and Purucci, O.,** DNA synthesis during the division cycle of three substrains of *Escherichia coli*, B/r, *J. Mol. Biol.,* 102, 477, 1978.
13. **Koppes, L. J. H., Woedringh, C. L., and Nanninga, N.,** Size variations and correlation of different cell cycle events in slow growing *Esherichia coli, J. Bacteriol.,* 134, 423, 1978.
14. **Kubitschek, H. E. and Freedman, M. L.,** Chromosome replication during the division cycle in slowly growing, steady state cultures of three *Escherichia coli* B/r strains, *J. Bacteriol.,* 136, 179, 1978.
15. **Kubitschek, H. E. and Newman, C. N.,** Chromosome replication during the division cycle in slowly growing steady-state cultures of three *Escherichia coli* B/r strains, *J. Bacteriol.,* 136, 179, 1978.
16. **Kubitschek, H. E.,Bendigkeit, H. G., and Loken, M. R.,** Onset of DNA synthesis during the cell cycle in chemostat cultures, *Proc. Natl. Acad. Sci., U.S.A.,* 57, 1611, 1967.
17. **Hiriyanna, K. T. and Ramakrishnan, T.,** DNA replication in *Mycobacterium tuberculosis* $H_{37}R_v$, presented at 15th Int. Congr. Genet., New Delhi, December 12 to 21, 1983.
18. **Youmans, G. P. and Karlson, A. G.,** Streptomycin resistance of tubercle bacilli. Studies on recently isolated tubercle bacilli and the development of resistance to streptomycin *in vivo, Am. Rev. Tuberc. Pulm. Dis.,* 55, 529, 1947.
19. **Barksdale, L. and Kim, K. S.,** Mycobacterium, *Bacteriol. Rev.,* 41, 217, 1977.
20. **Bradley, S. G.,** Reassociation of deoxyribonucleic acid from selected Mycobacteria with that from *Mycobacterium bovis* and *Mycobacterium farcinica, Am. Rev. Resp. Dis.,* 106, 122, 1972.
21. **Kornberg, A.,** *DNA replication,* W. H. Freeman & Co., San Francisco, 1980, 210.
22. **Cutter, R. G. and Evans, J. E.,** Synchronisation of bacteria by a stationary-phase method, *J. Bacteriol.,* 91, 469, 1966.
23. **Bird, R. and Lark, K. G.,** Initiation and termination of DNA replication after amino acid starvation of *Escherichia coli* $15T^-$, *Cold Spring Harbor Symp. Quant. Biol.,* 33, 799, 1968.
24. **Urban, J. E. and Lark, K. G.,** DNA replication in *Escherichia coli* $15T^-$ growing at 20° C, *J. Mol. Biol.,* 58, 711, 1971.
25. **Bird, R. and Lark, K. G.,** Chromosome replication in *Escherichia coli* $15T^-$ at different growth rates: rate of replication in the chromosome and the rate of formation of small pieces, *J. Mol. Biol.,* 49, 343, 1970.
26. **Lundquist, R., Manlapaz-Fernandez, P., and Olivera, B. M.,** A specific effect of deoxycitidine 5′-triphosphate on DNA chain elongation rate *in vitro, J. Mol. Biol.,* 83, 541, 1974.
27. **Olivera, B. M. and Bonhoeffer, F.,** Discontinuous DNA replication *in vitro* I. Two distinct classes of intermediates, *Nature New Biol.,* 240, 233, 1972.
28. **Rodriguez, R. L. and Davern, C. I.,** Direction of deoxyribonucleic acid replication in *Escherichia coli* under various conditions of cell growth, *J. Bacteriol.,* 125, 346, 1976.

29. **Abe, M. and Tomizawa, J.,** Chromosome replication in *Escherichia coli* affected in the process of DNA initiation, *Genetics,* 69, 1, 1971.
30. **Chandler, M. G. and Pritchard, R. H.,** The effect of gene concentration and relative dosage on gene output in *Escherichia coli, Mol. Gen. Genet.,* 138, 127, 1975.
31. **Beacham, I., Barth, P. T., and Pritchard, R. H.,** Constitutivity of thymidine phosphorylase in deoxy riboaldolase negative strains: dependance on thymine requirement and concentration, *Biochim. Biophys. Acta.,* 116, 589, 1968.
32. **Zaritsky, A., and Pritchard, R. H.,** Replication time of chromosome in thymineless mutants of *Esherichia coli, J. Mol. Biol.,* 60, 65, 1971.
33. **Maaloe, O. and Kjeldgaard, A.,** *Control of Macromolecular synthesis,* W. A. Benjamin, Inc., New York, 1966, 154.
34. **Bremer, H., Young, R., and Churchward, G.,** Initiation and termination of deoxyribonucleic acid replication in bacteria after a stepwise increase in the velocity of replication, *J. Bacteriol.,* 130, 92, 1977.
35. **Bremer, H. and Churchward, G.,** Deoxyribonucleic acid synthesis after inhibition of initiation of rounds of replication in *Escherichia coli* B/r, *J. Bacteriol.,* 130, 692, 1977.
36. **Churchward, G. and Bremer, G.,** Determination of deoxyribonucleic acid replication time in exponentially growing *Escherichia coli* B/r, *J. Bacteriol.,* 130, 1206, 1977.
37. **Chandler, M., Bird, R. E., and Caro, L.,** The replication time of *Escherichia coli* K12 chromosome as a function of the cell doubling time, *J. Mol. Biol.,* 94, 127, 1975.
38. **Churchward, G., Estwa, E., and Bremer, H.,** Growth rate dependency control of chromosome replication initiation in *Escherichia coli, J. Bacteriol.,* 145, 1232, 1981.
39. **Pritchard, R. H. and Zaritsky, A.,** Effect of thymine concentration on the replication velocity of DNA in a thymineless mutant of *Escherichia coli, Nature,(London)* 226, 126, 1970.
40. **MeMurrough, I. and Rose, A. H.,** Effect of growth rate and substrate limitation on the compositon and structure of the cell wall of *Saccharomyces cerevisiae, Biochem. J.,* 105, 189, 1967.
41. **Wehr, C. T. and Parks, W.,** Macromolecular synthesis in *Saccharomyces cerevisiae* in different growth media, *J. Bacteriol.,* 98, 458, 1969.
42. **Leick, V.,** Growth rate dependency of protein and nucleic acid composition of *Tetrahymena pyrifomis* and the control of synthesis of ribosomal and transfer RNA, *C. R. Trav. Lab. Carsberg.,* 36, 113, 1967.
43. **Cook, J. R.,** Adaptations in growth and division in *Euglena* affected by energy supply, *J. Protozool.,* 10, 436, 1963.
44. **Poyton, R. O.,** Effect of growth rate on the macromolecular composition of *Prototheca zopfii* a colorless alga which divides by multiple fission, *J. Bacteriol.,* 113, 203, 1973.
45. **Rose, T. K. and Yanofsky, C.,** Metabolic regulation of the tryptophan operon of *Escherichia coli*: repressor-independent regulation of transcription initiation frequency, *J. Mol. Biol.,* 69, 103, 1972.
46. **Koch, A.,** Overall controls of the biosynthesis of ribosomes in growing bacteria, *J. Theor. Biol.,* 28, 203, 1970.
47. **Bremer, H. and Dalbow, D. G.,** Regulatory state of ribosomal genes and physiological changes in the concentration of free ribonucleic acid polymerase in *Escherichia coli, Biochem. J.,* 150, 9, 1975.
48. **Churchward, G., Bremer, H., and Young, R. Y.,** Transcription in bacteria at different DNA concentrations, *J. Bacteriol.,* 150, 572, 1982.
49. **Manor, H., Goodman, D., and Stent, G. S.,** RNA chain growth rates in *Escherichia coli, J. Mol. Biol.,* 39, 1, 1969.
50. **Dougan, A. H. and Glaser, D. A.,** Rate of chain elongation of ribosomal RNA molecules in *Escherichia coli, J. Mol. Biol.,* 87, 775, 1974.
51. **Bremer, H., Hymes, J., and Dennis, P. P.,** Ribosomal RNA chain growth rate and RNA labelling patterns in *Escherichia coli* B/r, *J. Theor. Biol.,* 45, 379, 1974.
52. **Dennis, P. P., and Bremer, H.,** Regulation of ribonucleic acid synthesis in *Escherichia coli* B/r: an analysis of a shift-up. I. Ribosomal RNA chain rates, *J. Mol. Biol.,* 75, 145, 1973.
53. **Bremer, H., Barry, L., and Dennis, P. P.,** Regulation of ribonucleic acid synthesis in *Escherichia coli* B/r: an analysis of a shift-up. II. Fraction of RNA polymerase engaged in the synthesis of stable RNA at different steady-state growth rates, *J. Mol. Biol.,* 75, 161, 1973.
54. **Dennis, P. P. and Bremer, H.,** Regulation of ribonucleic acid synthesis in *Escherichia coli* B/r: an analysis of a shift-up. III. Stable RNA synthesis rate and ribosomal RNA chain growth rate following a shift-up, *J. Mol. Biol.,* 89, 233, 1974.
55. **Ryals, J., Little, R., and Bremer, H.,** Temperature dependence on RNA synthesis parameters in *Escherichia coli, J. Bacteriol.,* 151, 8, 79, 1982.
56. **Stent, G. and Bremer, S.,** A genetic locus for the regulation of ribonucleic acid synthesis, *Proc. Natl. Acad. Sci. U.S.A.,* 47, 2005, 1961.
57. **Gallant, J., Palmer, L., and Pao, C.,** Anomalous synthesis of ppGpp in growing cells, *Cell,* 11, 181, 1977.

58. **Travers, A., Kari, C., and Mace, H.,** Transcriptional regulation by bacterial RNA polymerase, in *Genetics as a Tool in Microbiology,* Grover, S. W. and Hopwood D. A., Eds., Cambridge University Press, London, 1981, 113.
59. **Harshey, R. M., and Ramakrishnan, T.,** Rate of ribonucleic acid chain growth in *Mycobacterium tuberculosis* $H_{37}R_v$ *J. Bacteriol.,* 129, 616, 1977.
60. **Burton, K.,** Analysis of RNA turnover in bacteria using histidine as a radioactivity trap for [2-^{3}H] adenine nucleotides, *J. Mole. Biol.,* 102, 333, 1976.
61. **Hershey, A. D.,** Conservation of nucleic acids during bacteriol growth, *J. Gen. Physiol.,* 38, 145, 1954.
62. **Chaney, S. G., and Boyer, P. D.,** Incorporation of water oxygens into intracellular nucleotides and RNA. II. Predominantly hydrolytic RNA turnover in *Escherichia coli, J. Mol. Biol.,* 64, 581, 1972.
63. **Norris, T. E. and Koch, A. L.,** Effect of growth rate on the relative rates of synthesis of messenger, ribosomal and transfer RNA in *Escherichia coli, J. Mol. Biol.,* 64, 633, 1972.
64. **Witkin, S. S. and Gibson, K. D.,** Changes in ribonucleic acid turnover during aerobic and anaerobic growth in *Rhodopseudomonas spheroides, J. Bacteriol.,* 110, 677, 1972.
65. **Witkin, S. S. and Gibson, K. D.,** Ribonucleic acid from aerobically and anaerobically grown *Rhodopseudomonas spheroides*: comparison by hybridization to chromosomal and satellite deoxyribonucleic acid, *J. Bacteriol.,* 110, 684, 1972.
66. **Fan, D. P., Higa, A., and Levinthal, C.,** Messenger RNA decay and protection, *J. Mol. Bio.,* 8, 210, 1964.
67. **Charnetzky, W. T. and Brubaker, R. R.,** RNA synthesis in *Yersinia pestis* during growth restriction in calcium deficient medium, *J. Bacteriol.,* 149, 1089, 1982.
68. **Maaloe, O.,** An analysis of bacterial growth, *Dev. Biol.,* 3(Suppl.), 33, 1969.
69. **Maaloe, O. and Kjeldgaard, N. O.,** Control of macromolecular synthesis: a study of DNA, RNA, and protein synthesis in bacteria, 1966, 210.
70. **Haseltine, W. A. and Block, R.,** Synthesis of ppGpp and ppGpp requires the presence of codon specific, uncharged tRNA in the acceptor site of ribosomes, *Proc. Natl. Acad. Sci. U.S.A.,* 70, 1564, 1973.
71. **Haseltine, W. A., Block, R., Gilbert, W., and Weber, K.,** MS I and MS II made on ribosome in idling step of protein synthesis, *Nature (London),* 238, 381, 1972.
72. **Cashel, M.,** The control of ribonucleic acid synthesis in *Escherichia coli, J. Biol. Chem.,* 244, 3133, 1969.
73. **Fiil, N., von Meyenberg, K., and Fresen, J. D.,** Accumulation and turnover of guanosine tetraphosphate in *Escherichia coli, J. Mol. Biol.,* 71, 769, 1973.
74. **Lindahl, L., Post, L., and Nomura, M.,** DNA-dependent *in vitro* synthesis of ribosomal proteins, protein elongation factors, and RNA polymerase subunit α: inhibition by ppGpp, *Cell,* 9, 439, 1976.
75. **Godson, G. N. and Sinsheimer, R. L.,** Use of Brij lysis as a general method to prepare polyribosomes from *Escherichia coli, Biochim. Biophys. Acta,* 149, 489, 1967.
76. **Harvey, R. J.,** Fraction of ribosomes synthesizing protein as a funcion of specific growth rate, *J. Bacteriol.,* 114, 287, 1973.
77. **Schlief, R.,** Control of production of ribosomal protein, *J. Mol. Biol.,* 27, 41, 1967.
78. **Van Dijk-Salkinoja, M. S. and Planta, R. J.,** Rate of ribosome production in *Bacillus licheniformis, J. Bacteriol.,* 105, 20, 1971.
79. **Varrichio, F. and Monier, R.,** Ribosome patterns in *Escherichia coli* growing at various rates, *J. Bacteriol.,* 108, 105, 1971.
80. **Michaelis, G. A.,** Ribosome maturation of *Escherichia coli* growing at different growth rates, *J. Bacteriol.,* 110, 889, 1972.
81. **Skhold, A. C., Juarez, H., and Hedgcoth, C.,** Relationships among deoxyribonucleic acid, ribonucleic acid, and specific transfer ribonucleic acids in *Escherichia coli* 15T$^-$ at various growth rates, *J. Bacteriol.,* 115, 177, 1973.
82. **Wong, J. T., Mustard, M., and Herbert, I.,** Regulation of amino acid-specific transfer RNAs in *Escherichia coli, Biochim. Biophys. Acta,* 174, 513, 1969.
83. **Williamson, D. H.,** The timing of deoxyribonucleic acid synthesis in the cell cycle of *Saccharomyces cerevisiae, J. Cell Biol.,* 25, 517, 1965.
84. **Sebastian, J., Mian, F., and Halvorson, H. O.,** Effect of the growth rate on the level of the DNA-dependent RNA polymerases in *Saccharomyces cerevisiae, FEBS Lett.,* 34, 159, 1973.
85. **Waldron, C. and Lacroute, F.,** Effect of growth rate on the amounts of ribosomal and transfer ribonucleic acids in yeast, *J. Bacteriol.,* 122, 855, 1975.
86. **Leick, V.,** Ratios between contents of DNA, RNA and protein in different microorganisms as a function of maximal growth rate, *Nature (London),* 217, 1153, 1968.
87. **Poyton, R. O. and Branton, D.,** The control of daughter-cell number variation in multiple fission: genetic versus environmental determinants in *Prototheca, Proc. Natl. Acad. Sci. U.S.A.,* 69, 2346, 1972.

88. **Pitale, M. P. and Jayaraman, R.,** Transcription of bacteriophage T4 genome *in vitro*. Heterogeneity of RNA polymerase in crude extracts of normal and T4 infected *Escherichia coli B, Biochemistry,* 14, 1265, 1975.
89. **Harshey, R. M. and Ramakrishnan, T.,** Heterogeneity of RNA polymerase in *Mycobacterium tuberculosis* $H_{37}R_v$, *J. Ind. Inst. Sci.*, 58, 557, 1976.
90. **Blatti, S. P., Ingles, C. J., Lindell, T. J., Morris, P. W., Weaver, R. F., Weinberg, F., and Rutter, W. J.,** Structure and regulatory properties of eucaryotic RNA polymerase, *Cold Spring Harbor Symp. Quant. Biol.*, 35, 649, 1970.
91. **Roeder, R. G.,** Eukaryotic nuclear RNA polymerases, in *RNA Polymerase*, Losick, R. and Chamberlin, M., Eds., Cold Spring Harbor Laboratory, Cold Spring Harbor, N.Y., 1976, 285.
92. **Dalbow, D. G.,** Synthesis of RNA polymerase in *Escherichia coli* B/r growing at different rates, *J. Mol. Biol.*, 75, 181, 1973.
93. **Matzura, H., Malin, S., and Maaloe, O.,** Sequential biosynthesis of the β and β′ subunits of the DNA dependent RNA polymerase from *Escherichia coli, J. Mol. Biol.*, 59, 17, 1971.
94. **Linn, T. and Scaife, J.,** Identification of a single promoter in *Escherichia coli* for rpl J, rpl L and rpo BC, *Nature (London)*, 276, 33, 1978.
95. **Newmann, A. J., Linn, T. G., and Haywood, R. S.,** Evidence for the cotranscription of the RNA polymerase genes rpo BC with ribosomal protein gene of *Escherichia coli, Mol. Gen. Genet.*, 169, 195, 1978.
96. **Bremer, H.,** Parameters affecting the synthesis of ribosomes and RNA polymerase in bacteria, *J. Theor. Biol.*, 53, 115, 1975.
97. **Iwakura, Y., Ito, K. and Ishihama, A.,** Biosynthesis of RNA polymerase in *Escherichia coli*. I. Control of RNA polymerase content at various growth rates, *Mol. Gen. Genet.*, 133, 1, 1974.
98. **Iwakura, Y. and Ishihama, A.,** Biosynthesis of RNA polymerase in *Escherichia coli*, II. Control of RNA polymerase synthesis during nutritional shift-up, *Mol. Gen. Genet.*, 142, 67, 1975.
99. **Kawakami, K., Saito, T., and Ishihama, A.,** Biosynthesis of RNA polymerase in *Escherichia coli*. IX. Growth-dependent variations in the synthesis rate, content and distribution of RNA polymerase, *Mol. Gen. Genet.*, 174, 107, 1979.
100. **Enami, M. and Ishihama, A.,** Biosynthesis of RNA polymerase in *Escherichia coli*. XII. Non-coordinate synthesis of core enzyme subunits after suppression of cell growth, *Mol. Gen. Genet.*, 185, 373, 1982.
101. **Engback, F., Gross, C., and Burgess, R. R.,** Quantitation of RNA polymerase subunits in *Escherichia coli* during exponential growth and after bacteriophage T4 infection, *Mol. Gen. Genet.*, 143, 291, 1973.
102. **Matzura, H., Hansen, B. S., and Zeuthen, J.,** Biosynthesis of the β and β′ subunits of RNA polymerase in *Escherichia coli, J. Mol. Biol.*, 74, 9, 1973.
103. **Shepherd, N. S., Churchward, G., and Bremer, H.,** Synthesis and activity of ribonucleic acid polymerase in *Escherichia coli, J. Bacteriol.*, 141, 1098, 1980.
104. **Burgess, R. R.,** Separation and characterization of the subunits of ribonucleic acid polymerase, *J. Biol. Chem.*, 244, 6168, 1969.
105. **Hayward, R. S. and Fyfe, S.,** Over-synthesis and instability of sigma protein in a merodiploid strain of *Escherichia coli, Mol. Gen. Genet.*, 159, 89, 1978.
106. **Pederson, S., Bloch, P. L., Reeh, S., and Neidhardt, F. C.,** Patterns of protein synthesis in *E. coli*: a catalog of the amount of 140 individual proteins at different growth rates, *Cell,* 14, 179, 1978.
107. **Chamberlin, M. J.,** Interaction of RNA polymerase with the DNA template, in *RNA Polymerase*, Losick, R. and Chamberlin, M., Eds., Cold Spring Harbor Laboratory, Cold Spring Harbor, N.Y., 1976, 159.
108. **Shepherd, N., Churchward, G., and Bremer, H.,** Synthesis and function of ribonucleic acid polymerase and ribosomes in *Escherichia coli* B/r after a nutritional shift-up, *J. Bacteriol.*, 143, 1332, 1980.
109. **Libby, R. T. and Brown, L. R.,** Ribonucleic acid polymerase subunit biosynthesis in *Bacillus subtilis, Mol. Gen. Genet.*, 185, 339, 1982.
110. **Ny, T. and Bjork, G. R.,** Growth rate-dependent regulation of transfer ribonucleic acid (5-methyluridine) methyl transferase in *Escherichia coli* B/r, *J. Bacteriol.*, 141, 67, 1980.
111. **Rich, A. and Raj Bhandary, U. L.,** Transfer RNA: molecular structure, sequence and properties, *Annu. Rev. Biochem.*, 45, 805, 1976.
112. **Bjork, G. R. and Kjellin-Straby, K.,** *Escherichia coli* mutants with defects in the biosynthesis of 5-methylaminomethyl-2-thiouridine or 1-methylguanosine in their tRNA, *J. Bacteriol.*, 133, 508, 1978.
113. **Burton, Z. F., Gross, C. A., Watanabe, K. K., and Burgess, R. R.,** The operon that encodes the sigma subunit of RNA polymerase also encodes ribosomal protein S21 and DNA primase in *Escherichia coli* K12, *Cell,* 32, 335, 1983.
114. **Nierlich, D. P.,** Regulation of bacterial growth, RNA and protein synthesis, *Annu. Rev. Microbiol.*, 32, 393, 1978.
115. **Mendelson, N. H.,** Bacterial growth and division: Genes, structures, forces and clocks, *Microbiol. Rev.*, 46, 341, 1982.
116. **Rohl, R.,** The biosynthesis of bacterial ribosome and its regulation, *Naturwissenschaften*, 70, 288, 1983.

Chapter 3

THE SPATIAL ORGANIZATION OF SECONDARY METABOLISM IN MICROBIAL AND PLANT CELLS

Werner Roos and Martin Luckner

TABLE OF CONTENTS

I. INTRODUCTION

The phylogenetically ancient (''primary'') reactions of metabolism resemble each other in all groups of organisms. During evolution they have been highly optimized and unified because of their prime importance for the elementary processes of life (assimilation of basic substrates, energy conversion, formation and degradation of proteins and nucleic acids, as well as of their precursors, etc.). In addition to these basic metabolic fields, a multitude of metabolic pathways exists which show a very sporadic distribution within the kingdom of living beings. They are characteristic of individual groups of organisms or of a few specialized cells and hence cannot be of significance for life in general. These reaction sequences are collectively called ''secondary metabolism'' and the compounds formed are called ''secondary metabolites''.

In contrast to the relatively small number of primary metabolites, which serve as their precursors, secondary products show an extreme diversity of molecular structures, as seen from the thousands of alkaloids, polyketides, isoprenoids, flavonoids, etc. occurring in nature. This biosynthetic variability has been conserved during evolution because it was only under low negative selection pressure (secondary pathways do not significantly lower the fitness of the primary metabolism) and/or attracted even a positive selection pressure: many secondary compounds have acquired useful functions in the ecologic strategy and the individual development of the producing organism.

At the cellular level, the co-existence of primary and secondary pathways requires that the following main problems be solved:

- Secondary pathways have to compete with reactons of primary metabolism for common substrates (amino acids, shikimic acid, acetyl-CoA, sugars, etc.) as well as energy sources and cofactors.
- Essential cellular functions have to be protected from the damaging influence of some secondary compounds.
- Many secondary metabolites are required to serve as signal molecules within cellular differentiation processes.

Hence, in order to maintain or to increase the competitiveness of the producing cell or organism, secondary metabolism must be precisely coordinated with various other cellular functions. This ''tuning'' occurs both in space and in time and involves different levels of regulation. They comprise:

1. The amount and the kinetic properties of the enzymes of secondary pathways
2. The temporal coordination of the synthesis and degradation of the secondary metabolic enzymes within programs of cellular differentiation
3. The integration of enzymes, precursors, intermediates, and products into the spatial organization of the living cell

Topics (1) and (2) have been intensively discussed in recent reviews.[1-4] This paper is devoted to the structural aspects (3). They are seen as the background of the vectorial functioning of secondary pathways and are discussed on the basis of some of the best investigated plant and microbial systems.

II. POOLS OF PRECURSORS AND INTERMEDIATES

The initial step of each secondary pathway is the branching from a primary metabolic chain resulting in the competition for a common precursor. The key metabolites feeding both primary and secondary enzymic sequences are frequently sequestered between distinct cellular compartments. Such "pools" are established either within membrane-surrounded organelles or in the microenvironment of protein molecules or membrane surfaces. The concentration of the pooled molecules is regulated via their rates of membrane transfer (influx and efflux) or via the extent of their noncovalent binding, respectively. Both parameters may reflect the rates of biosynthesis and degradation and the influences of various regulatory mechanisms including catalytic transport activity, noncatalytic membrane permeability, cooperative binding, etc. Such mechanisms are also involved in the cellular control of the exchange of metabolites between different pools. In each case, equilibration of pooled metabolites due to random diffusion is largely reduced.

Frequently from the compartmentation of a common precursor directly originates the spatial separation of the following reaction sequences: different pools may serve selectively as reservoirs of primary and secondary enzymic chains located in their close vicinity. This is, however, not generally the case. Up to now, no pools of primary metabolites are known which show an exclusive destination to secondary metabolism. It seems rather that preexisting pools of primary compounds serve both primary and secondary pathways, i.e., in many cases different enzymic chains are supplied by the same precursor pool. The spatial separation of primary and secondary pathways (and, of course, also of reaction sequences within each field of metabolism) may then be realized at another stage of structural organization (spatial arrangement of enzymes (see Section III); channeling of intermediates (see Section IV); compartmentation of products (see Section V).

A. Examples

1. Distribution of Amino Acids in Eukaryotic Cells

In fungal[5-11] and plant[12-14] cells, free amino acids are generally distributed within at least three compartments: the cytosolic pool, the mitochondrial and/or chloroplast pool, and a nonoplasmatic storage pool. The controlled exchange between these compartments gives rise to characteristic concentration gradients of amino acids in eukaryotic cells. In the first two pools, a minimum concentration is maintained for the supply of protein synthesis and the various amino acid converting and degrading enzymes localized therein. The storage capacity of these metabolically active compartments is usually low; excess amino acids are accumulated in a nonplasmatic, metabolically inactive pool. The latter is localized either in the vacuole (as shown with yeast,[5,6,15] *Penicillium cyclopium*,[17] and cells of higher plants [13,14]) or in vesicles of similar function (as shown with *Neurospora crassa*).[10] The vacuolar (vesicular) amino acids serve as a reservoir which compensates the influence of changes of the external or biosynthetic amino acid supply on the content of the cytosolic amino acid pool: during nitrogen starvation a net efflux into the cytosol takes place; on the other hand,

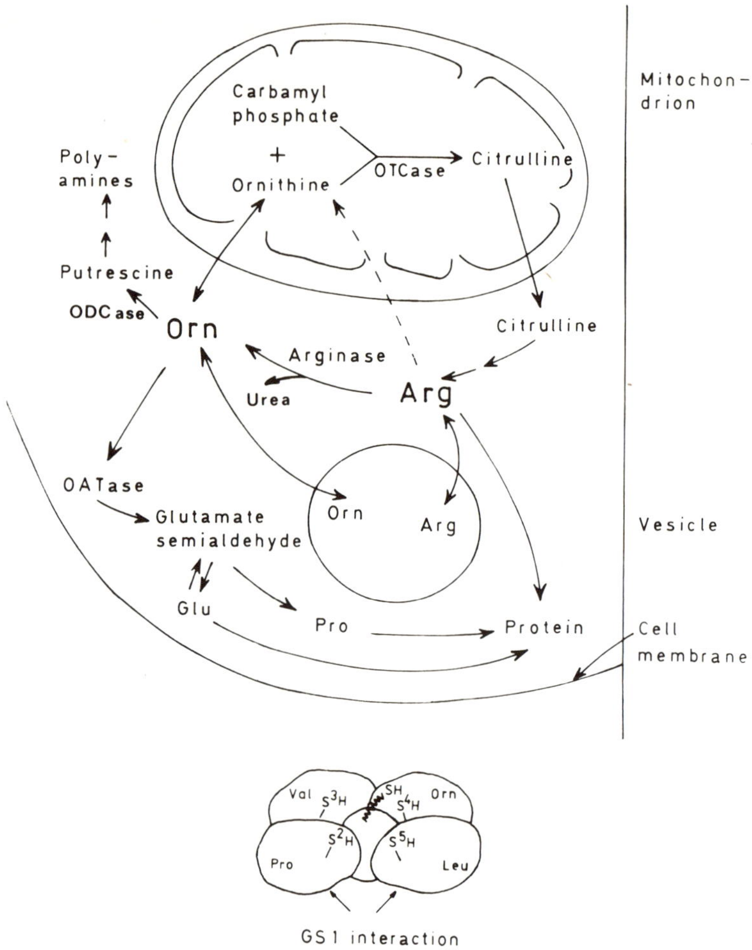

FIGURE 1. Ornithine metabolism in *Neurospora crassa*.[9-11] OTCase, ornithine transcarbamylase; OATase, ornithine aminotransferase; ODCase, ornithine decarboxylase. See text for explanation.

externally added amino acids are channeled via the cytosol to the vacuoles and accumulated there up to several thousand-fold. The ability of the cell to maintain such high concentration differences between the cytosol and the vacuole constitutes the high storage capacity of this compartment. The underlying mechanism is not generally understood. In *Saccharomyces cerevisiae*, the vacuolar arginine seems to be trapped by polyphosphates.[16] In *Penicillium cyclopium*, the accumulation of various amino acids is due to a dynamic equilibrium between active vacuolar uptake and passive reflux into the cytoplasm, resulting in a recycling of intracellular amino acids across the vacuolar membrane.[17,18]

2. Compartmentation and Spatial Separation of Metabolic Pathways of Ornithine in Neurospora crassa

In this organism, the relationship between the aforementioned amino acid pools and individual metabolic sequences has been extensively studied with the metabolism of L-ornithine (Figure 1).[9-11]

1. In the mitochondria, where L-ornithine is synthesized, it can be reacted with carbamylphosphate by ornithine transcarbamylase (OTC). The L-citrulline formed is an intermediate in the biosynthesis of L-arginine.

2. In the cytosolic pool, L-ornithine is accessible to two enzymes. Decarboxylation by ornithine decarboxylase (ODC) to putrescine initiates a sequence leading to the polyamines spermine and spermidine, whereas the formation of glutamic-γ-semialdehyde by ornithine aminotransferase (OAT) opens the path to either L-glutamic acid or L-proline.
3. Excess L-ornithine is withdrawn from enzymic attack by accumulation in intracellular vesicles.

The contents of the last two pools exchange readily with the cytosolic L-ornithine. In cells growing exponentially on a minimal medium, a steady state is reached with approximately 1% of total L-ornithine in the cytosol, 3% in the mitochondria, and more than 95% in the vesicles. The relatively low stationary concentration of cytosolic ornithine (about 0.125 m*M*) allows nearly the maximum conversion rate of the high-affinity enzyme ODCase, but is far below the saturating concentration required by the low-affinity OATase. Hence, the synthesis of polyamines is limited by the capacity of the key enzyme, whereas the formation of proline is controlled by the cytosolic concentration of L-ornithine. Consequently, during growth in minimal medium, both pathways drain only 8 and 11%, respectively, of the total L-ornithine synthesized. An increase of the cytosolic ornithine concentration has less effect on the rate of polyamine synthesis, but enhances severalfold the rate of proline formation. (The cytosolic concentration of L-ornithine can be raised by feeding either extracellular L-ornithine or L-arginine. The latter amino acid, which is one of the products of L-ornithine metabolism, causes several regulatory effects: (1) feedback inhibition of the mitochondrial biosynthesis of L-ornithine; (2) inhibition of uptake of L-ornithine into the mitochondria, and hence inhibition of resynthesis of arginine; and (3) increased uptake of cytosolic L-arginine into the vesicles and concurrent efflux of vesicular L-ornithine, possibly by exchange via a common carrier. L-Ornithine made from cytosolic L-arginine by urease is not stored but degraded via the proline path.

Summarizing, in this example the secondary pathway (polyamine synthesis) is deeply embedded into the primary metabolic grid. It is held at a low, but nearly constant rate, whereas the rates of primary reactions respond to environmental signals by both enzymic and compartmental regulation.

3. Attribution of L-Phenylalanine Pools to Alkaloid and Protein Biosynthesis in Penicillium cyclopium

In this fungus, L-phenylalanine serves as a precursor of the peptide alkaloids cyclopenin and cyclopenol.[19-21] Isotopic dilution analysis suggests that the cytoplasmic as well as the vacuolar pool of this amino acid feed the biosynthesis of the alkaloids, whereas protein synthesis is supplied by cytosolic L-phenylalanine only. Thus, labeled external L-phenylalanine may reach the site(s) of alkaloid biosynthesis directly via the cytosolic pool (giving rise to "primary labeling" of the alkaloids) or indirectly after accumulation in the vacuolar pool (causing "secondary labeling"). It must be noted, however, that under all experimental conditions used, external L-phenylalanine contributes by no more than 10% to the total alkaloid synthesis, i.e., the main supply of this precursor comes either from the *de novo* synthesis of L-phenylalanine or the degradation of protein.[22]

III. LOCATION OF SECONDARY METABOLIC ENZYMES

A. Distribution Between Different Organelles

Enzymes of secondary metabolism occur in all plasmatic and nonplasmatic compartments (cf. Luckner et al.[23] for a recent review). In particular, plastids, microsomes, Golgi vesicles, microbodies, the cytosol, and extraplasmic space were shown to contain secondary metabolic enzymes.

Under the aspect of metabolic compartmentation, two main principles can be distinguished. Frequently, enzymes belonging to different pathways which compete for a common pool of precursor or intermediate are located in different organelles, i.e., the interorganellar distribution of enzymes reflects the spatial separation of metabolic branches.

In other cases, different organelles contain similar enzyme activities, whose products there may feed different metabolic sequences. Thus, some early enzymic steps common to various secondary pathways are present in parallel in several organelles; branching and spatial separation may then occur at later, more specialized steps within an organelle (see Section IV). The multiplication of basic enzymic steps is frequently indicated by the occurrence of isoenzymes. In some cases, the presence of a distinct isoform in one particular organelle could be demonstrated, emphasizing the metabolic individuality of that compartment. The interorganellar distribution of enzymes and their connections with distinct pools of precursors or intermediates span a complex spatial network of primary and secondary reactions, which may further be nested by compartmentation within organelles. At present, no complete picture of such a network can be outlined, but the main elements are emerging from an increasing number of details reported from different objects and classes of secondary compounds.

B. Example

1. Metabolism of Phenylpropanoid Phenolics in Higher Plants (Figure 2; for reviews, see References 24 and 25)

The key molecules of this metabolic field are the CoA-esters of cinnamic acid and its -OH and -OCH_3 derivatives. They are derived from L-phenylalanine via the so-called general phenylpropanoid metabolism. This pathway is initiated by phenylalanine ammonia-lyase (PAL). All enzymes involved have been found to be associated at least in part with microsomes, suggesting that the ER membranes bear the complete reaction sequence. Consecutive enzymes may be closely coupled, as shown with PAL and the enzymes hydroxylating its product cinnamic acid[25a] (see Section IV.C.3). The cinnamic acid CoA-esters produced in the ER are transferred to and further transformed in other organelles:

1. In the chloroplasts, caffeoyl-CoA is a substrate of chlorogenate synthase, which catalyzes esterification with quinic acid to chlorogenic acid. In the same organelle, *p*-coumaroyl-CoA is converted by chalcone synthase to naringenin chalcone, which is further processed to the various flavonoids.[26]
2. Feruloyl- and sinapoyl-CoA most probably become encapsulated within secretory vesicles derived from the ER and/or the Golgi apparatus.[27,135] These vesicles are suggested to contain cinnamoyl-CoA reductases and cinnamoyl alcohol dehydrogenases.[28] These enzymes catalyze the formation of cinnamoyl alcohols, which are monomers of lignins.[29] The vesicles merge with the cytoplasmic membrane and excrete their content into the extraplasmic space. During the transport of the vesicles, di- and oligomerization of the cinnamic alcohols are initiated, whereas the final polymerization to lignins proceeds within the cell wall.[28,30] The polymerization is catalyzed by peroxidases (with the aid of an H_2O_2-generating enzyme system) which occur both in cellular vesicles and (after excretion) bound to the cell wall.[28,30-32]
3. A special type of secretory vesicle containing presumably no converting enzymes carries cinnamoyl-CoA esters to the cell wall where they are used in the formation of the ester moieties found in different lignins.[33,34]

In addition to this branched pathway, which is realized by interorganellar cooperation, there exist other reaction sequences starting from L-phenylalanine, which proceed *in toto* within distinct organelles. A frequent characteristic of such "autarkic" phenylpropanoid

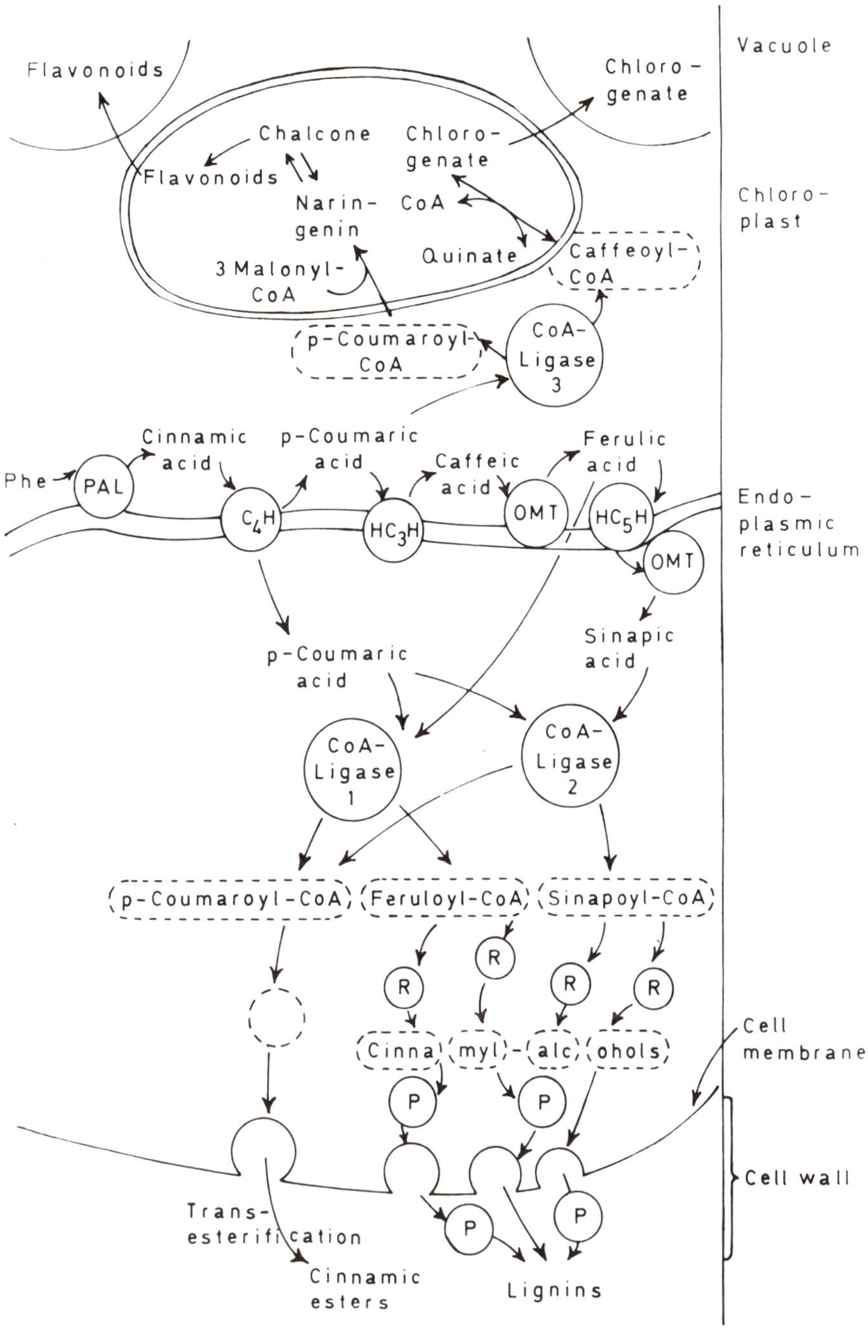

FIGURE 2. The distribution of some metabolic activities involved in the metabolism of phenylpropanoid phenolics in plant cells.[24] PAL, phenylalanine ammonia-lyase; HC_4H, cinnamate 4-monooxygenase; HC_3H: p-coumarate 3-monooxygenase OMT, catechol O-methyltransferase; HC_5H, ferulate 5-hydroxylase (hypothetical); R, cinnamoyl-CoA:NADPH oxidoreductases and cinnamoyl alcohol dehydrogenases; P, peroxidases. See text for explanation.

pathways is the occurrence of complexes between PAL and distinct enzymes catalyzing the conversion of its product cinnamic acid:

1. In chloroplasts of higher plants, PAL is associated with cinnamate 2-monooxygenase and further enzymes leading to the formation of coumarins[35,36] and furanocoumarins,[37] but is not involved in the synthesis of chlorogenic acid and flavonoids. (As described above, the formation of chlorogenic acid requires cinnamate 4-monooxygenase which, with one exception,[38] is exclusively located in the ER membranes.[24,36]) The last step in the formation of the methoxylated furanocoumarins, however, requires most probably the release of intermediates from the chloroplasts and their transfer to another site of further processing, e.g.,*O*-methylation.[37]
2. In different organelles, benzoic acid and *p*-hydroxybenzoic acid are formed by β-oxidation of the 3-carbon side chain of cinnamic acid or (after 4-hydroxylation) of *p*-coumaric acid. Complexes of PAL with the enzymes involved have been found in the thylakoid membranes of chloroplasts of higher plants[36,39,40] and green algae[40] (synthesis of *p*-hydroxybenzoic acid), in the mitochondrial inner membrane[41] (formation of *p*-hydroxybenzoic acid), in the chromophore membranes of blue-green algae[42] (formation of benzoic acid and *p*-hydroxybenzoic acid), and in nonphotosynthetic organs of *Ricinus communis* in glyoxysomes[43] (formation of benzoic acid).

The selective involvement of organelles in the various phenylpropanoid pathways is further indicated by the occurrence of organelle-specific isoforms of PAL. For example, two PAL isoenzymes with different sensitivity to feedback inhibition occur in various plants. In *Quercus pedunculata*, one of them (which can be inhibited by cinnamic acids) cosediments with the microsomal fraction and cinnamate 4-monooxygenase activity, whereas the other form (which is sensitive to benzoic acids) is associated with the mitochondria-microbody fraction and with the activity of the benzoic acid synthetase system.[44,45] In the glandular tissue of *Aesculus*, one isoform of PAL was localized in the ER and another in the plastids.[46] The latter one is part of a plastidal enzyme complex producing benzoic acids[39] (see above). Up to now it is not clear whether isoenzymes involved in secondary pathways are products of different genes or are derived by post-translational modifications of a single enzyme.

C. Membrane Association of Enzymes

There is evidence for many enzymes of secondary metabolism to be either integrated in or bound to cellular membranes.[23] The experimental proof usually involves the demonstration of enzyme activity in isolated and purified membrane fractions obtained from cell homogenates (mostly by density gradient centrifugation). Though elaborate procedures exist,[47,47a] the production of artifacts in membrane isolation cannot always be excluded. In the homogenates, soluble enzymes may bind to carbohydrate moieties of membranes; on the other hand, weak associations of enzymes to membrane surfaces often do not survive the disintegration procedures. Hence, enzymes found in a soluble fraction might well be associated with membranes in vivo. Data concerning the membrane association of enzymes should, therefore, not only be derived from cell fractionation studies and microscopical visualization, but also make use of the following characteristics of membrane-related enzymes.

1. Binding of and Interaction with Membrane Components

In many cases, enzyme activities can be solubilized from membrane fractions. The procedure required for the enzyme in question may indicate a preferred interaction either with apolar domains or with ionic groups of the membrane. The first case may be exemplified with the membrane-embedded cinnamate 4-monooxygenase, which requires detergents for solubilization,[48] and the latter case with catechol *O*-methyltransferase, which is removed from ER membranes in the presence of $(NH_4)_2SO_4$ and EDTA.[49] Stabile binding to membrane lipids may be resistant to detergents as shown with cyclopenase, a lipoprotein of the cytoplasmic membrane of the conidiospores of *Penicillium cyclopium*.[50] The dependence on (or modulation by) membrane lipids of the activity of purified membrane proteins is well known for many enzymes of primary metabolic pathways. Up to now, similar phenomena were scarcely studied with secondary enzymes. An example is the C_{55} isoprenoid alcohol kinase from *Staphylococcus aureus*.[51-53]

2. Nature of the Substrate

The substrates of membrane-embedded enzymes are usually more lipophilic, especially less hydroxylated than those of surface or soluble enzymes. In the biosynthesis of gibberellins in *Cucurbita maxima*, for instance, the nonpolar intermediates between ent-kaurene and GA_{12}-aldehyde are converted by membrane-bound enzymes, whereas the following, more polar intermediates are substrates of soluble enzymes.[54,55]

Enzymes involved in hydroxylations appear to be generally membrane bound (e.g., monooxygenases of the general phenylpropanoid metabolism or of the biosynthesis of gibberellins). Other groups are usually soluble, e.g., esterifying and transmethylating enzymes.

3. Vectorial Functioning of the Enzyme In Vivo

The binding of an enzyme at a membrane surface or its integration into the membrane structure determines its spatial orientation and therefore provides the basis for the vectorial character of the catalyzed reaction. This may manifest itself as confined accessibility to substrates or cofactors, etc., as membrane transfer of the substrate or product, or as cooperation with other enzymes or transport proteins arranged at the same membrane. Phenomena of this type are discussed in the next chapter.

IV. CHANNELING OF INTERMEDIATES

A. General Principles

The cellular arrangement of metabolite pools, catalytic and binding sites, constitutes spatially defined routes along which metabolites are to be transferred following distinct reaction sequences. Usually the intermediates moving on such paths are in part or totally prevented from undergoing a diffusional equilibrium with othe regions of the cell. In this way, mixing with the metabolites of other (even competing) chains is greatly reduced or abolished, i.e., a metabolic channel is established. It allows the predominant or exclusive routing of pooled precursors (or of substrates coming from outside the cell) to distinct metabolic destinations. The selectivity as well as the catalytic efficiency of a channeled sequence essentially depends on the degree to which random diffusion of the intermediates is replaced by their controlled movement in the sequence of catalytic sites. The main principles of cellular diffusion control which give rise to characteristic channeling phenomena are aligned with the two levels of cellular compartmentation:

- In the cellular macrospace, membranes serving as diffusion barriers keep the intermediates of a sequence restricted to an organelle. On the other hand, specific transmembrane channels (e.g., carrier systems) allow metabolic cooperation between different compartments (see Section IV.B).
- In the cellular microspace, the diffusional path (and hence the diffusion time) between binding or catalytic sites may be minimized by the formation of enzyme aggregates. The microenvironment of these protein complexes further restricts the diffusional "escape" of intermediates. In this way, microcompartments are constituted in the cytoplasm, within plasmatic organelles, or at their membranes. They enable short-range channeling at protein or membrane surfaces (see Section IV.C).

B. Transfer across Membranes

Due to the widespread involvement of membranes in the cellular compartmentation of precursors, intermediates, and enzymes, crossing of membrane barriers is a necessary step in many secondary metabolic sequences. As membrane transfer of metabolites is often rate limiting, it represents a potential site of regulation of metabolic fluxes. The basic mechanisms involved in the membrane transfer of secondary metabolites fall into four categories:

1. Penetration of low polar compounds by noncatalytic diffusion across the lipid domains of membranes (usually found also for lipophilic xenobiotics; (see Section IV.B.1)
2. Carrier-catalyzed transport (see Section IV.B.2)
3. Coupling of transfer with chemical transformation catalyzed by membrane-embedded enzymes with vectorial function (see Section V.B.1)
4. Secretion of metabolites into membrane vesicles via (2) or (3), followed by the transfer of these vesicles; after merging with the target membrane, the content of the vesicles is set free at the *trans*-side of that membrane (see Sections III.B.1 and V.B.2).

1. Noncatalytic Membrane Permeability

The substrate supply of many secondary enzymes is controlled by the noncatalytic permeability of membrane barriers. A typical example is cyclopenase, an enzyme which is tightly bound to the cytoplasmic membrane of the conidiospores of *Penicillium cyclopium*.[50] Surface cultures of this fungus excrete the benzodiazepine alkaloids cyclopenin and cyclopenol into the culture liquid. These alkaloids are substrates of cyclopenase, which catalyzes their conversion to the quinoline derivatives viridicatin and viridicatol.[21] The conversion rate of extracellular cyclopenin and cyclopenol is negligibly small in normal cultures or suspensions of native conidiospores, but increases by about two orders of magnitude after membrane disruption by X-Press or acetone treatment. It is suggested, therefore, that cyclopenase is located at the plasmatic side of the plasmalemma whose permeability strongly limits the substrate supply.[58] The noncatalytic permeability of cellular membranes may be increased by reducing the ATP content of the cells below a critical level, e.g., by treatment with azide.[58,56] Under such conditions, the benzodiazepine alkaloids present in the medium of surface cultures of *Penicillium cyclopium* become accessible to cyclopenase and are completely converted.[57] In this permeabilized state, the alkaloid metabolism of this strain is shifted towards the patterns normally found in strains of *Penicillium viridicatum*, which, after a phase of excretion of cyclopenin/cyclopenol into the medium, take up these alkaloids and convert them to viridicatin/viridicatol.[57]

2. Regulation of Secondary Pathways by Means of Membrane Permeability

The noncatalytic and catalytic transport properties of biological membranes are subject to temporal changes which are correlated with the programs of cellular differentiation. Such changes, which result from modifications in the patterns of membrane constituents, are of special importance for metabolic cooperation between cellular organelles. An example is the transient permeability increase of the envelope of etioplasts during their light-induced transformation into chloroplasts.[59-61] Within a distinct period, before the onset of CO_2 fixation, transport of some metabolites across the plastid envelope is remarkably enhanced. Among the molecules which are preferentially transferred are primary metabolites, some of them being potential precursors of secondary products, e.g., succinate, acetate, and oxaloacetate, key intermediates of secondary pathways, e.g., mevalonate, and CoA-esters of cinnamic acids, and secondary products, such as abscisic and gibberellic acids. Many of the substances transferred into the plastids originate from the mitochondria; the inner membrane of these organelles exhibits a similar permeability increase, which appears to be synchronized with the changes of the plastid envelopes.[59,62,63] The physiological significance of this developmental flux control lies in the supply of plastidal biosynthetic pathways with various precursors, which is necessary as long as plastids are unable to fix CO_2.

Whereas the primary metabolites among the aforementioned transportants are known to be substrates of specific carrier systems of the inner chloroplast membrane,[64,65] transfer of the secondary compounds most probably depends on the noncatalytic permeability of this membrane. It is assumed, therefore, that a reorganization of the inner membrane takes place during maturation of the plastids. This process must be of specific nature as (1) other metabolites (glycine, α-oxoglutarate, glutamate, glutamine) show less or no increase in transport rates and (2) the malate/mevalonate exchange system (Section IV.B.3) remains uninfluenced by the transient raise of the transport rates.[61] During maturation of the chloroplasts, specific changes proceed in the peptide pattern of their envelopes.[66,67] The underlying molecular mechanism is not yet known. Mediation by photochrome has been suggested, as this photosensor was found at the envelope of etioplasts and within the plastids.[68]

The modulation of membrane permeability by plant hormones, e.g., ethylene or phytochrome, is well known from higher plants. Ethylene increases, for instance, the permeability of the tonoplast in cells of *Ipomoea* flowers, thereby causing loss of turgor and initiating various metabolic reactions, e.g., those leading to the fading of the corolla.[69] It has been suggested that the increased permeability of the tonoplast enhances the accessiblity of the ethylene precursor L-methionine accumulated in the vacuoles to the cellular site(s) of ethylene formation.[70,71] This may explain the known autocatalytic induction of ethylene biosynthesis during senescence of the *Ipomoea* flowers.[72]

The phytochrome system mediates a variety of light effects on secondary metabolism. The molecular mechanisms are best understood with the light-induced synthesis of anthocyanins, i.e., widespread flower pigments of phenolic structure. The biologically active form of phytochrome formed by red light irradiation initiates not only the formation of the anthocyanin-synthesizing enzymes,[73-75] but also the reversible changes of membrane permeability,[76] which control the substrate supply of these enzymes. Reagents which increase the noncatalytic permeability of membranes, e.g., *n*-propanol, DMSO, DMF, and Ca^{2+} chelators, stimulate the biosynthesis of anthocyanins in a manner resembling active phytochrome.[77,78] On the contrary, membrane-stabilizing agents like cholesterol or Ca^{2+} inhibit the increase of the rate of anthocyanin synthesis, irrespective of whether it was triggered by active phytochrome or *n*-propanol.[79]

The influence of phytochrome on membrane permeability probably rests on the reversible liberation of membrane-bound Ca^{2+} as indicated by an increase of free Ca^{2+} concomitant with the red light activation of phytochrome.[79] (Ca^{2+} ions are of basic importance for the modulation of the lipid and protein arrangement in biological membranes.) Phytochrome may further affect the metabolism of membrane-bound molecules (e.g., gibberellin-like substances[80,81]), as well as the activity of membrane-bound proteins.[76] The phytochrome-mediated control of membrane permeability may produce specific regulatory effects. An intriguing example is the influence of light on the qualitative composition of lignins in poplar stems.[25] Under conditions yielding a high level of active phytochrome, isolated xylem tissues form lignin containing syringyl and guajacyl residues, whereas only syringyl residues are incorporated into lignin synthesized in the presence of low levels of active phytochrome. Membrane permeabilizers, e.g., *n*-propanol, or membrane stabilizers, e.g., Ca^{2+}, give rise to similar effects. These results probably reflect the co-existence of permeability-controlled and -uncontrolled enzymic reactions in lignin biosynthesis. It is tempting to speculate that among the known CoA ligase isoenzymes involved in the formation of lignin precursors,[25,29,82,83] the one responsible for the activation of ferulic acid is separated from its substrate by a phytochrome-controlled membrane barrier. Other possible reasons for selective membrane effects are yet to be investigated, e.g., different Ca^{2+} sensitivities of membrane barriers or the compartmentation of active phytochrome.[84]

3. Carrier-Mediated Transport of Secondary Metabolites

In contrast to primary metabolism, evidence for the involvement of carrier systems in the transport of secondary intermediates is very rare. One exception is the transport of mevalonate across the envelope of chloroplasts, which occurs in exchange with pre-accumulated malate and vice versa.[61] Uptake of one or both substances proceeds on account of an equimolar efflux of the other. This coupling has a relatively high specificity (mevalonate can be replaced with decreasing efficiency by oxaloacetate, α-oxoglutarate, and glutamate, but not by acetate), indicating the existence of a mevalonate/malate exchange carrier in the inner chloroplast membrane.

C. Channeling within Microcompartments

1. Kinetic Properties of Enzyme Aggregates

This section deals with the short-range channeling of metabolites between enzymes associated within functional complexes. Irrespective of the great variability of their molecular architecture and of the cooperation between the catalytic units, all of these microcompartments are characterized by the limitation or absence of free diffusion of their intermediates.[85] This is the basis of a common diagnostic feature called "catalytic facilitation", i.e., the preferred or exclusive conversion of a precursor moving via the coupled sequence compared with similar molecules coming from outside the channel. This general kinetic property of a metabolic channel can most easily be demonstrated with isolated microcompartments. The experimental proof should ascertain the following:

- The initial substrate of the channel gives rise to a higher overall conversion rate than does any intermediate.
- At suboptimal concentrations of the initial substrate, addition of intermediates does not (or does not to the expected degree) increase the rate of product formation.
- After simultaneous application of differently labeled initial substrate and intermediate, the isotope ratio of the end product shows a higher share of the label coming from the initial substrate than does the isotope ratio of the intermediate.

The last-mentioned approach has proved to be an especially powerful tool in demonstrating (as well as determining the degree of) coupling in enzyme complexes.[25a,38,40,86]

The unique catalytic properties of an enzyme complex enable the coordinate control of activity of the assembled enzymes. For instance, in the arom complex — a multifunctional polypeptide in the prearomatic pathway of *Neurospora* — preincubation with the initial substrate causes a reduction in the lag phase of the overall reaction[87] and a coordinate activation of the following enzymes.[88] Generally, the V_{max} of the second reaction is higher than that of the first step, which is to be expected in order to prevent saturation of the second step and accumulation of intermediates.

As will be shown, microcompartments provide the following main advantages for the metabolic regulation of secondary pathways:[89,90]

- An increased catalytic efficiency due to reduced times required for the transfer of intermediates and for reaching the steady state of product formation
- The segregation of competing metabolic sequences by means of vectorial transfer of precursors
- The possibility of a common metabolic control of the associated enzymes (coordinate activation or inhibition, changes in the sensitivity towards the microenvironment)
- The protection of chemically unstable intermediates
- The protection of the surrounding cytoplasm from potentially toxic effects of intermediates or products

2. Soluble Multienzyme Complexes

Within the plasmatic organelles and the cytosol, enzymes of many metabolic chains are joined within functional aggregates. The degree of binding of the catalytic units varies widely, ranging from very loose to more tight noncovalent associations of individual enzymes, and culminates in covalent linkages between the catalytic domains in polyfunctional proteins. In each case, the metabolic chain attains a vectorial character, the spatial arrangement of catalytic sites determining the route of intermediates and hence the sequence of enzyme reactions.

The diffusional "escape" of intermediates from such channels may be reduced or prevented in two ways:

- The precise orientation and concerted action of the coupled enzymes allow the intermediates to pass from one binding position to the next without being in a free state during the whole course of reactions. Usually a special carrier protein is involved in the intrinsic transfer of the bound metabolites.
- Intermediates dissociate from the catalytic sites after each conversion step; however, due to the specific microenvironment of the protein complex (restricted mobility of water, proximity of functional groups interacting by ionic and/or van der Waals forces), diffusion is confined to a distinct region surrounding the array of catalytic sites.

Regarding structural organization, three principal types of soluble microcompartments can be distinguished, and are discussed next.

a. Loose Enzyme Aggregates Capable of Reversible Association-Dissociation

In the last years, many classical "soluble" enzymic sequences were suggested to operate in vivo as weakly bound aggregates. Typical examples are the glycolytic chain[91] or the coupling of mitochondrial malate dehydrogenase and citrate synthase.[92] As such structures are generally dissociated during cell disintegration, evidence for their existence rests mainly on the demonstration of catalytic facilitation in intact cells or organelles and of interactions between purified enzymes under conditions mimicking a normal cellular milieu. Enzymes interacting near the association-dissociation equilibrium may be of special regulatory importance because of their self-aggregating properties. The actual degree of aggregation — which modifies their catalytic properties — may be under cellular and environmental control.

Cooperation between such loosely associated enzymes usually includes reversible, noncovalent binding and (limited) diffusion of free intermediates. (Exceptions are the sets of fatty acid-synthesizing enzymes in some bacteria and in chloroplasts.[93]) In some cases, transfer of intermediates between individual enzymes may be facilitated by reversible binding to noncatalytic binding proteins. Examples are the sterol binding proteins found in animal cells[94,95] and the (suggested) kaurene binding protein of pea seeds.[96] The latter protein most probably translocates kaurene synthesized by "soluble" enzymes to enzymes present in the ER membranes.

b. Stable Enzyme Complexes Formed by Noncovalent Binding

Complexes of this type are practically completely in the associated state under cellular conditions, although the binding strength varies widely. At least some of them survive mild cell disintegration procedures, so their catalytic behavior may be studied in vitro. The individual subunits disaggregate by treatment with SDS or other protein dissociating agents. In many cases, among the subunits are multifunctional polypeptides that are either identical or that complement each other.

- Identical subunits each containing the complete set of catalytic sites are found, e.g., in the 6-methylsalicylic acid synthetase of *Penicillium patulum*.[97,98]

- Subunits containing complementary parts of the enzymic chain are found in the anthranilate synthetase of *Neurospora crassa*[99] and the fatty acid synthetase of yeast.[100] From the field of secondary metabolism, the synthetases involved in the formation of polyketides and in the nonribosomal synthesis of many antibiotic peptides belong to this group.[101] The interaction of the subunits is not only required to realize the complete reaction sequence, but sometimes may also cause an allosteric activation of catalytic sites. This was shown both with complexes consisting of individual enzymes (e.g., tryptophane synthase[99]) and those containing multienzymes (e.g., gramicidine S synthetase;[102] see below).

c. *Multifunctional Proteins (Polyenzymes)*

The covalent linkage of several catalytic units within a single polypeptide chain represents the most stable form of a metabolic channel. The usual evidence for such polyenzymes is the demonstration of a multistep catalysis by a purified protein, which is stable against strong protein-dissociating agents, e.g., SDS at 90°C. Although the detailed molecular structure of a polyfunctional peptide is yet to be elucidated, some principles of the cooperation between functional domains have been established in the best investigated system (see Section IV.C.2.c.1). All known polyenzymes of secondary metabolism are involved in a distinct group of reactions: the oligomerization of small carboxylic acids (acetic, propionic, malonic, methylmalonic) or amino acids to fatty acids, polyketides, or peptides, respectively. Catalysis occurs at protein templates preloaded with the monomers in the proper sequence. The "template capacity" of a single polyfunctional enzyme was found to be six units at maximum. For instance, synthesis of the antibiotic peptide enniatin B (linkage of six amino acids) and of flavanones (addition of three acetate units to a hydroxylated cinnamyl-CoA molecule) is catalyzed by one polypeptide chain. Due to this limitation, the formation of larger molecules is achieved by the noncovalent aggregation of several polyfunctional polypeptide chains. The largest molecule known to be synthesized in this way, the antibiotic peptide alamethicin, consists of 20 amino acids.

Polyenzymes as well as their noncovalent complexes operate with covalently bound intermediates, the binding step being directly involved in the chemical activation of the building blocks. A special catalytic unit containing 4-phosphopantetheine mediates the "handing over" of metabolites between the catalytic sites via the reversible formation of thioesters. This "carrier protein" plays a central role for the structural and functional organization of protein templates, as shown by its widespread occurrence, irrespective of whether the subunits are only loosely aggregated (*E. coli*, fatty acid synthetase[93]), tightly but noncovalently bound (*Bacillus brevis*, tyrocidine synthetase[103]), partly polyfunctional (yeast fatty acid synthetase[93]), or fully polyfunctional (*B. brevis*, gramicidine S synthetase 2[104]).

A polyenzyme functioning without a 4-phosphopantetheine-containing subunit is the chalcone synthase isolated from parsley cells.[105] The three acetate units which are condensed by this enzyme with 4-coumaroyl-CoA are derived from malonyl-CoA by decarboxylation, as is usually the case with acetate-condensing polyenzymes. In contrast to other enzymes (e.g., fatty acid synthases), the intermediate acetylcarbanions most probably react as CoA-ester and not in an enzyme-bound form.

1. Example: Formation of Gramicidine S in *B. brevis* ATCC 9999[101,102,104]

The biosynthesis of this decapeptide (cyclo-[D-Phe-Pro-Val-Orn-Leu]$_2$) occurs on a soluble multienzyme complex consisting of two multifunctional polypeptide chains. Gramicidine-synthetase 1 (GS 1; 100,000 daltons) bears the initiation site, where the starter amino acid L-Phe is activated as an aminoacyladenylate, and then is bound as a thioester and epimerized to D-Phe. Likewise, the other substrates (Pro, Val, Orn, Leu) are AMP activated and then are bound in this sequence to thiol groups of the polyenzyme GS 2 (280,000 daltons), which serves as a "thiotemplate" for the following steps.

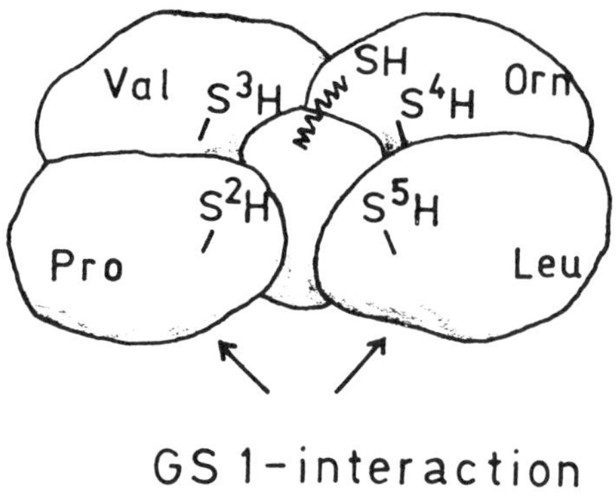

FIGURE 3. Model of the GS 2 multienzyme of gramicidin S-synthetase.[101,104] The subunits responsible for the activation of the indicated amino acids are arranged around the pantetheine-carrying subunit. See text for explanation.

Synthesis is initiated by the transfer of D-Phe from GS 1 to a specific binding site on GS 2 from where it attacks the NH-group of the Pro-thioester to form the (thio-activated) dipeptide *S*-Pro-D-Phe. This peptide is then transferred to the SH-group of a centrally arranged 4′-phosphopantetheine-residue of GS 2. The newly formed thioester undergoes a stepwise "intrinsic" transfer to attack with its NH_2-terminal the thio-activated carboxyl of the proximal amino acid, this giving rise to the formation of an enzyme-bound pentapeptide, which, by head-to-tail-cyclization with a second molecule, yields gramicidine S.

Each activation and elongation step is considered to be catalyzed by a functional domain containing a binding site for the AMP-amino acid and a peripheral SH-group. Attempts to isolate the individual amino acid-activating units by limited proteolysis were only partially successful, as proteolytic cleavage occurred both between and within the functional domains. Nevertheless, a detailed analysis of proteolytic fragments activating several amino acids as well as mutational alterations of the enzyme structure show that the sequence of activating units corresponds to the amino acid sequence of gramicidine S. The polyenzymes TY 2 and TY 3 involved in the formation of the similar antibiotic tyrocidine are disintegrated by an (unidentified) proteolytic factor from *B. brevis* to amino acid-activating units, each of 70,000 daltons.[106] These findings led to a structural model of peptide-forming polyenzymes with similar-sized amino acid-activating domains arranged around a 4-phosphopantetheine-containing "peptidyl carrier protein" (PCP) (see Figure 3).

The initiation of peptide synthesis requires the formation of a complex of the heavy enzyme GS 2 with GS 1. This complex allows not only the transfer of D-Phe (see above), but establishes cooperativity between both enzymes, as shown by the fact that synthesis of gramicidine S with the heavy enzyme loaded with all five substrate amino acids requires the presence of the light enzyme.[102]

3. Coordination of Enzymes at Membrane Surfaces

Biological membranes frequently serve as a matrix for the assembly of enzymes involved in a metabolic chain. The precise orientation of their catalytic sites in close proximity (which is realized by both protein-lipid and protein-protein interaction) mediates the channeling of intermediates along the surface of the membrane, but also the cooperativity between individual enzyme activities. Usually, intermediates are released from the catalytic sites, but are prevented from random diffusion by the low permeability of the membrane barrier and the microenvironment of the protein-membrane complex. A special variant of such membrane-bound microcompartments is the association of metabolic enzymes with transmembrane channels, which allows the coordination of enzymic conversions with the membrane transfer of the end product (see Section V.B.1).

In the field of secondary metabolism, channeling within membrane-bound enzymic sequences has been observed most frequently between phenylalanine-ammonia lyase (PAL) and enzymes converting the intermediate cinnamic acid. Conversion of phenylalanine and tyrosine into *o*- and *p*-coumaric acid or benzoic acid and *p*-hydroxybenzoic acid is catalyzed not only by photosynthetic membranes prepared from chromophores of blue-green algae,[42] thylakoids of algae,[38] and chloroplasts,[39] but also by mitochondrial inner membranes[41] and microsomes.[25a] In all of these preparations, the cinnamic acid formed intermediary was shown not to be in equilibrium with externally added cinnamic acid. The degree of coupling between PAL and the cinnamic acid-converting enzymes (measured as catalytic facilitation; see Section IV.B) varies between two extremes: it is low in the formation of *p*-coumaric acid from phenylalanine on ER membranes or on thylakoids, and is very tight in the benzoate synthetases in the chromophore membranes of blue-green algae or of chloroplasts and in the *p*-hydroxybenzoate synthetase in mitochondria. In the latter cases, externally added intermediates are practically not accepted as a substrate.[41]

The interaction of membrane-bound enzymes essentially depends on the integrity of the membrane matrix. With ER membranes converting phenylalalnine into *o*- and *p*-coumaric acids, the degree of catalytic facilitation was greatly reduced during membrane purification or treatment with light or with ethylene.[25a]

a. Example: A Four-Step Channel in the Formation of p*-Hydroxymandelic Acid Nitrile in Sorghum bicolor*

A microsomal membrane preparation obtained from germinating seedlings of *S. bicolor* catalyzes the conversion of L-tyrosine to *p*-hydroxymandelonitrile by the following reaction sequence:[107]

$$\text{L-Tyrosine} \xrightarrow{(O)} N\text{-hydroxy-tyrosine} \xrightarrow[CO_2]{-2H} p\text{-hydroxyphenyl-acetaldoxime}$$

$$\xrightarrow{-H_2O} p\text{-hydroxyphenyl-acetonitrile} \xrightarrow{(O)} p\text{-hydroxy-mandelonitrile}$$

(The end product is further converted to the cyanogenic glucoside dhurrin by a soluble enzyme.) By two sets of experiments, catalytic facilitation has been demonstrated within this chain, suggesting that the membrane-bound enzyme system constitutes a metabolic channel:[86,107,108]

- Although each labeled intermediate is incorporated into the mandelonitrile, the rate of end product formation is significantly higher with tyrosine and *N*-hydroxytyrosine compared with the late intermediates, i.e., the aldoxime and the acetonitrile.
- ^{3}H-aldoxime and ^{3}H-acetonitrile produced from ^{3}H-tyrosine do not readily exchange with added ^{14}C-aldoxime or ^{14}C-acetonitrile, but instead are preferentially incorporated into the end product.

V. SECRETION AND STORAGE OF SECONDARY PRODUCTS

A. Significance of Compartmentation of Secondary Compounds

The products of secondary pathways are usually stored apart from the centers of metabolic activity. The transfer to and accumulation in nonplasmic regions is one of the protective strategies of the cell against the disturbing or toxic effects of these compounds on metabolism.

The potential toxicity of their own secondary products for many plant and microbial cells becomes apparent when the breakdown of compartmentation or channeling allows the contact of these molecules with regions of vital metabolic importance, like enzymes or cellular membranes. The following examples are chosen from a variety of similar observations:

- The tannins, which are produced by many plant cells, are strong protein denaturing agents. In growing cells, the cytosolic concentration of these phenolics is held at a tolerable level by their accumulation in the vacuole[109,110] and by an equilibrium of synthesis and remetabolization.[111] An increase of cytosolic tannins (caused by, e.g., a damage of the tonoplast) leads to cellular death, as shown for instance in late phases of development of cell suspension cultures.[112]
- Soybean cells produce the phytoalexin glyceolin by the so-called "hypersensitive response reaction" towards plant pathogens.[113] This compound exerts a site-specific inhibition of mitochondrial electron transport not only at potential microbial invaders, but also at mitochondria isolated from the same plant.[114] Similarly, the flavonol kaempferol produced in many plant species causes both inhibition of respiration and electron flow when added to intact plant mitochondria.[115]
- The toxicity of many secondary compounds is due to their disturbing effects on the molecular architecture of cellular membranes. In living cells, the membrane contact of such molecules may be reduced by trapping (see below) or by channeling within the plasmic phase. Furthermore, the sidedness as well as the asymmetric distribution of the attackable membrane components implicates different sensitivites of both membrane sides towards specific membrane-active compounds. This may explain why cells or organelles tolerate the presence or accumulation of such molecules at one face of a membrane, but get damaged if treated with the same agent from the other face. For instance, among the various isoquinoline alkaloids present in the latex of *Chelidonium,* sanguinarine and chelerythrine are preferentially bound within the latex vacuoles. Both compounds cause lysis of isolated vacuoles if the added amount exceeds the accumulation capacity of these organelles.[116-117]
- Numerous phytoalexins, like pisatin,[118] rishitin,[119] or phaseollin,[120] cause a loss of selective permeability or even lysis of the cytoplasmic membrane if added to protoplasts isolated from tissues producing these compounds.
- In some cases, "harmless" secondary products which are stored in vacuoles of intact cells are converted into cytotoxic compounds when the breakdown of compartmentation allows their contact with degrading enzymes. In *Melilotus, Asperula,* and other plant species, *o*-coumaric acid glycosides are produced and subsequently stored in the vacuoles. Destruction of tonoplast and plasmalemma (e.g., during wilting or rupture of leaves) makes the glycosides accessible to β-glucosidases localized in the cell wall, thus giving rise to the immediate formation of the uncoupling agent — coumarin.[121] The glucosinolates accumulated in the vacuoles of the storage parenchyma of various Cruciferae and other plants are separated solely by the tonoplast from the cytosolic enzyme thioglucosidase; damage of this membrane initiates their conversion to the toxic isothiocyanates (mustard oils).[122] In other cases, secondary compounds and degrading enzymes are localized in different tissues; in *Sorghum bicolor,* the cyanogenic glycoside dhurrin is accumulated in the vacuoles of epidermal cells,[123] whereas the enzymes β-glucosidase and hydroxynitrilase (which may give rise to the liberation from dhurrin of HCN) are found in the cytoplasm and the chloroplast, respectively, of mesophyll cells.[124,125]

B. Principles and Mechanisms of Storage and Excretion of Secondary Products

Secondary products are stored both in plasmatic and nonplasmatic compartments of living cells. Strongly lipophilic compounds may form lipid droplets within the cytoplasm, i.e., they constitute a nonpolar phase immiscible with the cytosol. This has been shown for carotinoids,[126] resins, and essential oils in plant glands[127] and rubber particles.[128] In such droplets other lipophilic compounds may be solved, e.g., ergot alkaloids.[129] Most probably, lipophilic droplets exist also in vacuoles. For instance, in plant cells that produce anthocyanins, these pigments are frequently observed in an intensely colored globule, which originates from numerous smaller spherical structures and is finally located within the central vacuole. It was conjectured that this globule, which was termed anthocyanoplast, represents a membrane-bound organelle containing at least the late enzymes of anthocyanin biosynthesis.[130a,130b] The occurrence of a nonplasmatic, membrane-coated organelle within the nonplasmatic medium of the vacuole would, however, not be consistent with the hitherto existing experience according to which a single biomembrane can only be stable between a plasmatic and a nonplasmatic phase.[146] Furthermore, recent findings suggest the anthocyanoplasts to be hydrophobic droplets rather than membrane-surrounded vesicles.[130c]

Relatively small amounts of lipophilic secondary metabolites may be incorporated into different types of cellular membranes. The most prominent examples are carotenoids which occur in the cytoplasmic, ER, mitochondrial, and thylakoidal membranes of higher plants.[130,131] In chloroplast membranes of wheat, the specific incorporation of gibberellins has been demonstrated.[132]

By far the most part of secondary products is excreted from plasmatic sites of synthesis into nonplasmatic compartments:

- The vacuole obviously represents the dominant storage compartment of secondary products in higher plants. Among others, plant phenolics (condensed tannins, anthocyanins, flavonols, hydroxycinnamic acids) and precursors thereof (shikimic acid, esters of hydroxycinnamic acids), cyanogenic glucosides, glucosinolates, saponins, cardiac glycosides, and alkaloids have been localized in vacuoles of plant cells or in the isolated organelle (for reviews, see Matile[133] and Matile and Wiemken[134]).
- Various secondary products are excreted into the extracytoplasmic space. The most important examples are the constituents of the cell wall, like lignins,[27,29,135] cellulose,[136-138] and pectic polymers,[135,139-141] but also tannins[142] and sporopollenin.[143] As discussed below, excretion of the monomeric building blocks of the cell wall is tightly connected with polymerization reactions. Among the low molecular secondary metabolites found in the periplasmic space are chalcones, flavonoid glycosides,[144] and di- and triterpenoids.[127,142,145] Also, various enzymes involved in late steps of secondary product formation or degradation of secondary compounds have been localized in noncytoplasmic compartments.[30,30a] Hence, secondary products may also be completed or degraded outside the plasmatic phase. In the extracytoplasmic area, some enzymes are retained by the sieve effect of the cell wall network; others may covalently be bound to the cell wall.
- Plasmatic and nonplasmatic compartments of the cell are generally separated by biological membranes.[146] Therefore, removal of secondary products from the protoplasm involves membrane transfer of these compounds. The present knowledge about the secretion process is rather heterogeneous. This situation reflects the variability of mechanisms involved in the cellular transfer of secondary products, but also the restriction of experimental work to distinct aspects and objects, which was dictated by the available techniques. Despite the lack of information on the molecular level, the following main principles involved in secretion and storage of secondary products are apparent.

1. Coupling of Transfer with Chemical Conversion

The last biosynthetic step, which yields a product harmful to the cell, may be coupled with the membrane transfer of this product. The whole process is catalyzed by a membrane-embedded enzyme system with vectorial mode of action.

For instance, in *Bacillus brevis* Vm 4, the polyenzymes synthesizing the polypeptide edeine are associated with a complex of cytoplasmic membrane and DNA. It has been suggested that this spatial array permits the fast excretion of the antibiotic, thus avoiding its inhibitory effects on DNA and protein synthesis.[147] Another example is the synthesis of cellulose in most algae and in higher plants.[136,148] It occurs at a highly organized multisubunit enzyme complex embedded in the cytoplasmic membrane, which catalyzes polymerization of the D-glucan chains at the outer surface of that membrane. Ordered arrays of protein subunits, which are most probably identical with cellulose synthetase complexes, were observed by freeze-fracture studies within the cytoplasmic membrane of algae and higher plants. These "terminal complexes" are situated at the end of growing microfibrils. The almost simultaneous association by hydrogen bonds of the newly formed chains results in the formation of partially crystalline microfibrils. It has been suggested that the force created by the act of polymerization propels the complex through the fluid-mosaic membrane. The direction of motion, which determines the highly ordered orientation of the produced microfibrils, may be guided through the influence of microtubules.[148,149] Little is known about the mechanism of transport of the substrate (most probably UDP-glucose) to the site of polymerization at the outer surface of the membrane. Suggestive evidence indicates that protein- or lipid-linked intermediates (containing glucose or its dimer cellobiose) may be involved. An up to date model of cellulose biosynthesis is hypothesized by Delmer.[136]

Cytochemical observations frequently indicate the first appearance of secondary products within the lumen of the endoplasmic reticulum (ER) or the cisternae of the Golgi apparatus (GA). As the surrounding endomembranes of both systems — isolated as microsomal fractions of cell homogenates — bear a great number of secondary biosynthetic enzymes,[23] coupling of synthesis with the transfer across these membranes is very likely. In this way the vesicular excretion of the product may be originated.

2. Vesicular Transfer

Various secondary products and intermediates are compiled in secretory vesicles which are continuously generated by the endomembrane system (ER; GA). These vesicles represent specific nonplasmatic "transfer compartments" which selectively merge with the cytoplasmic or the vacuolar membrane. In this way their content is delivered either into the periplasmic space or the vacuole. Vesicular excretion across the cytoplasmic membrane (exocytosis) has been observed with the monomeric intermediates of lignin biosynthesis[135] (cinnamoylalcohols; see Section III.A), the matrix polysaccharides of the cell wall (pectine and hemicelluloses[139,140]), and slime polysaccharides.[141,142] The origin of the secretory vesicles involved in lignification has not yet been identified unequivocally. The synthesis of the mentioned polysaccharides clearly occurs within the dicytosomes of the Golgi apparatus from which they are transported with remarkable velocity to the periplasmic space. The average time of vesicle transfer is in the order of a few minutes.[151] In a small group of organisms, the haptophycean flagellates, even cellulose is transferred by vesicles: these algae produce a pseudowall consisting of several cellulosic scales which are synthesized via microfibrils within Golgi cisternae, and then rapidly secreted to the cell exterior.[137,138]

Vesicular transport into the vacuole is less documented by experimental data. The most conclusive evidence exists for the vacuolar deposition of vegetable tannins, i.e., polyphenolic compounds which can be localized electron microscopically, e.g., after fixation with caffeine.[109] It has been suggested from such studies that synthesis occurs at the rough and smooth ER, combined with the secretion of the product into the lumen and segregation of vesicles which are transferred to and fuse with the tonoplast, thereby releasing the tannins into the vacuole.[110-112,152] It was further conjectured that peroxidases present in the vacuole might catalyze the last biosynthetic step, i.e., the condensation of catechines to condensed tannins,[133] much like the function of extracellular peroxidases in lignification (see Section III.A).

The vesicular transfer of secondary products is part of the so-called flow differentiation of biomembranes, a fundamental process critical to growth, development, and homeostasis of living cells. It comprises a continuous flow of membrane elements from rough ER and nuclear envelope (generating elements) via smooth ER, Golgi apparatus, and related vesicles (transition elements), to plasma membrane or tonoplast (end products). Prior or during transit, membranes undergo a progressive differentiation by addition, deletion, or alteration of membrane components, i.e., they acquire the functional specialization typical for plasma membrane or tonoplast.[153] In addition, this process generates those specific surface properties of secretory vesicles which allow the selective recognition of their target membrane and subsequent fusion with it.[150] The movement of these vesicles is further directed by cytoskeletal structures of the cytosol.[154] Flow differentiation of membranes also involves acquisition or loss of secondary metabolic enzymes, as suggested from the specific distribution of such enzymes[23] and by metabolic conversions of secondary compounds during vesicle transfer (for instance, in lignin biosynthesis; see Section III.B.1).

3. Transfer by Cooperation between Organelles

An alternative (or complementary) way to the vesicular excretion of secondary products may be the direct association of the ER membranes with other cellular compartments. An impressive example is the production and export of flavonoids in glandular cells of buds of *Populus nigra*.[144] The formation of the flavonoids takes place in invaginations of the plastid envelope filled with cytoplasm containing mainly smooth ER (infraplastidal cytoplasm). Membrane transfer of flavonoids from ER into the plastid may be combined with their methylation. During the secretion process, close interconnections have been demonstrated between smooth ER and vacuome, as well as between ER and invaginations of the plasmalemma. In the nonplasmatic lumen of all of these organelles, temporary accumulation of flavonoids occurs. The observed continuity of cellular membranes suggests a thorough way of transport of the flavonoids from the site of synthesis to the vacuoles and to the extraplasmatic space by crossing closely connected membranes and hence minimal contact with the cytosol. Unfortunately, nothing is known about the molecular mechanisms underlying the membrane transfer of the flavonoids.

Little is known about the mechanism of excretion of lipophilic secondary products by plant glands, even though a great number of morphological observations are available.[127,142] They indicate a clear correlation between the synthesis of essential oils, resins, and other terpenoids, flavonoid aglycons, fats, etc., and the development of the smooth endoplasmic reticulum (but not of the Golgi apparatus). There are, however, no indications that ER-derived secretory vesicles carry these compounds to the exterior of the cells. Therefore, direct transfer through the cytoplasm of "harmless" compounds, followed by eccrine elimination, i.e., passive or active transport across the plasmalemma, has been conjectured.[127]

4. Trapping Mechanisms

The accumulation of secondary products in noncytoplasmic areas may reach considerable concentration gradients;[116,122,124] in many cases, removal of these compounds from the cytoplasmic phase is practically complete.[133] Up to now there is little evidence for active membrane transport of secondary compounds to be involved in this distribution. It seems rather that high local accumulations of secondary products are frequently due to their immobilization within nonplasmatic compartments. This is achieved by (1) conversion to derivatives which are unable to penetrate the separating membrane or (2) binding to insoluble cellular structures. The first case may be exemplified with the accumulation of alkaloids in the vacuoles of higher plant cells. Due to their hydrophobicity, the uncharged molecules passively penetrate the tonoplast, followed by their protonation in the acidic milieu of the vacuole. As the protonated species cannot repenetrate the tonoplast, diffusion of alkaloids into the vacuole proceeds unidirectionally as long as the vacuolar sap is more acidic than the cytosol. The slope of this pH gradient mainly determines the accumulation ratio of alkaloids between vacuole and cytoplasm as shown, e.g., by the distribution of nicotine.[155,156] Accordingly, in cells specialized for the vacuolar storage of alkaloids (which occur, e.g., in various Papaveraceae and Apocynaceae), the pH of the vacuolar sap was found to be markedly lower than in unspecialized cells.[157,159]

The acidification of the vacuole occurs by an H^+-pumping ATPase located in the tonoplast.[161-164] The generated H^+ gradient is not only required for the trapping of protonizable compounds, but especially for the uptake of ions and various primary metabolites by catalytic H^+ antiport.[165,166]

Trapping of alkaloids via protonation most probably also occurs within the intermembrane space of the chloroplast surface on account of the pH gradient produced by photosynthesis. This suggestion was derived from the high accumulation of nicotine observed in rooted leaves of *Nicotiana* and the localization of this alkaloid between the chloroplast membranes (and in infraplastidal vacuoles).[158-160]

Retention of alkaloids in vacuoles may also occur by binding of the protonated species to anionic compounds (cationic trapping). For instance, the isoquinoline alkaloids present in the latex vacuoles of *Chelidonium* seem at least in part to be bound to chelidonic acid and polyphenols. Interestingly, the binding sites show the highest affinity for the cytotoxic (tonoplast permeabilizing; see Section V.A) alkaloids sanguinarin and chelerythrin; the appearance of these compounds in the cytosol leads to their rapid uptake into vacuoles paralleled by the displacement of the prebound ''harmless'' alkaloids berberine and coptisine. Furthermore, in the course of vacuole development, synthesis of the aforementioned toxic alkaloids is preceded by the formation of the phenolic compounds involved in their binding.[116,117] Thus, the vacuolar trapping shows some specifity in favor of the most cytotoxic compounds. It should be mentioned that not in all cases could the vacuolar storage of alkaloids be attributed to simple trapping phenomena. For instance, accumulation of nicotine in vacuoles of *Nicotiana* seems to be somehow correlated with the electrical potential difference measured between the vacuole and the outer medium.[160]

The accumulation and retention of the various glycosides present in plant vacuoles are not yet understood. A trapping mechanism based on the glycosylation of the aglyca during or after tonoplast transfer (the increased hydrophilicity of the products would prevent their rediffusion) is less likely, since glycosylation in the cytosol has been repeatedly observed.[24,167] Tentative results point to a catalytic (or even active) transport of cytosolic glycosides via a vacuolar transport system which is under cytoplasmic control.[133]

Recently, the mechanism of vacuolar uptake of glycosides and alkaloids has been studied with isolated vacuoles in more detail. Barley mesophyll protoplasts convert the exogenously added coumarinaglycon esculetin to the glucoside esculin which is accumulated in the vacuole. Uptake of the hydrophilic glycoside by isolated vacuoles shows saturation kinetics, is stimulated by ATP, and inhibited by the protonophore CCCP. These results are interpreted as carrier-mediated transport across the tonoplast which is energized by the pH-gradient generated by the tonoplast ATPase.[168]

Experiments with vacuoles isolated from *Catharanthus roseus* cell cultures indicate that the indole alkaloids produced in the cytoplasm are concentrated in the vacuole by an energy-requiring mechanism which is sensitive to temperature and pH, stimulated by K^+ and Mg^{2+}, and blocked by ATPase inhibitors. Alkaloids pre-accumulated in the vacuoles can be released by subsequent addition of excess external alkaloid.[169] These findings do not necessarily prove the existence of an alkaloid specific carrier in the vacuolar membrane but may also be consistent with an ATPase-dependent proton trapping mechanism. In the latter case, once the protonation capacity of the vacuole is near saturation, incoming alkaloid should exchange with the protonated species and thus initiate their efflux. It was further shown that vacuoles isolated from different plant tissue cultures can accumulate only those alkaloids which are produced by the particular plant species.[169] This finding attests to a high degree of specificity of the uptake mechanism, which may either indicate an alkaloid specific vacuolar carrier system (as concluded by the authors) or selected permeability properties of the tonoplast.

Binding of cationic secondary products occurs to a significant extent by cell wall constituents which may act as cation exchanger. Examples are the strong absorption by xylem vessels of berberine and nicotine in *Berberis* and *Nicotiana*, respectively.[158,160]

VI. CONCLUSIONS

From the aspect of the producing cell, secondary metabolism is the result of co-evolution of secondary product formation and of the spatial organization of synthesis and storage of secondary compounds. The organization of secondary pathways in the cellular space enables not only a high efficiency of secondary biosynthesis and its regulation, but at the same time makes this production compatible with the requirements of primary metabolism. In particular, spatial organization:

- Economizes the consumption of precursors
- Regulates the competition and avoids mutual disturbance of primary and secondary metabolic chains
- Prevents the contact of sensitive metabolic areas with toxic secondary products

The main strategies involved in the spatial control of secondary metabolism comprise:

- The establishment of pools, i.e., areas of accumulation of precursors and intermediates
- The spatial distribution of secondary metabolic enzymes between and within cellular organelles
- The vectorial transfer (channeling) of precursors and intermediates, favoring distinct reaction sequences by (1) membrane transport and cooperation between organelles and (2) coupling of catalytic steps within microcompartments
- The controlled storage of secondary products, preferentially in nonplasmatic compartments, as well as the release and/or degradation of stored compounds

No compartments with an exclusive destination for secondary metabolism were found till now. Obviously, the pre-existing spatial organization with all its compartments, subcompartments, membranes, and molecular superstructures employed in the primary metabolism of the cell can be adapted to the needs of secondary product formation. This situation, which demonstrates the high flexibility in the basic construction of living cells, becomes especially clear by the involvement of secondary metabolism in basic processes of cellular functioning, such as the flow differentiation of biomembranes and the formation of and cooperation between organelles.

REFERENCES

1. **Luckner, M., Nover, L., and Böhm, H.,** Secondary metabolism and cell differentiation, in *Molecular Biology, Biochemistry and Biophysics,* Vol. 23, Springer-Verlag, Berlin, 1977.
2. **Martin, J. F. and Demain, A. L.,** Fungal development and metabolite formation, in *The Filamentous Fungi,* Vol. 3, Smith, J. E. and Berry, D. R., Eds., John Wiley & Sons, New York, 1978, 426.
3. **Luckner, M.,** Expression and control of secondary metabolism, in *Encyclopedia of Plant Physiology,* (New Series), Vol. 8, Bell, E. A. and Charlwood, B. V., Eds., Springer-Verlag, Berlin, 1980, 23.
4. **Wiermann, R.,** Secondary plant products and cell and tissue differentiation, in *The Biochemistry of Plants,* Vol. 7, Conn, E. E., Ed., Academic Press, New York, 1981, 85.
5. **Wiemken, A. and Nurse, P.,** Isolation and characterization of the amino acid pools located within the cytoplasm and vacuoles of *Candida utilis, Planta,* 109, 293, 1969.
6. **Wiemken, A. and Dürr, M.,** Characterization of amino acid pools in the vacuolar compartment of *Saccharomyces cerevisiae, Arch. Microbiol.,* 101, 45, 1974.
7. **Huber-Wälchli, V. and Wiemken, A.,** Differential extraction of soluble pools from the cytosol and the vacuoles of yeast (*Candida utilis*) using DEAE-dextran, *Arch. Microbiol.,* 120, 141, 1979.
8. **Schwencke, J. and de Robichon-Szulmajster,** The transport of *S*-adenosyl-L-methionine in isolated yeast vacuoles and spheroplasts, *Eur. J. Biochem.,* 65, 49, 1976.
9. **Weiss, R. L.,** Intracellular localization of ornithine and arginine pools in *Neurospora, J. Biol. Chem.,* 248, 5409, 1973.
10. **Davis, R. H., Bowman, B. J., and Weiss, R. L.,** Intracellular compartmentation and transport of metabolites, *J. Supramol. Struct.,* 9, 473, 1978.
11. **Bowman, B. J. and Davis, R.,** Cellular distribution of ornithine in *Neurospora:* anabolic and catabolic steady states, *J. Bacteriol.,* 130, 274, 1977.
12. **Oaks, A. and Bidwell, R. G. S.,** Compartmentation of intermediary metabolites, *Annu. Rev. Plant Physiol.,* 64, 88, 1979.
13. **Wagner, G. J.,** Content and vacuole/extravacuole distribution of neutral sugars, free amino acids and anthocyanins in protoplasts, *Plant Physiol.,* 64, 88, 1979.
14. **Boudet, A. M., Canut, H., and Alibert, G.,** Isolation and characterization of vacuoles from *Melilotus albus* mesophyll, *Plant Physiol.,* 68, 1354, 1981.
15. **Wiemken, A.,** Compartmentation and control of amino acid utilization in yeast, in *Cell Compartmentation and Metabolic Channeling,* Nover, L., Lynen, F., and Mothes, K., Eds., VEB Fischer, Jena, 1980.
16. **Dürr, M., Urech, H., Boller, Th., Wiemken, A., Schwencke, J., and Nagy, M.,** Sequestration of arginine by polyphosphate in vacuoles of yeast (*Saccharomyces cerevisiae*), *Arch. Microbiol.,* 121, 169, 1979.
17. **Roos, W. and Luckner, M.,** Enzymic and compartmental regulation of alkaloid biosynthesis in *Penicillium cyclopium,* in *Overproduction of Microbial Metabolites,* FEMS Symp. No. 13, Krumphanzl, V., Sikyta, B., and Vanek, Z., Eds., Academic Press, London, 1982, 111.
18. **Roos, W.,** Kinetik und Regulation des Aminosäure-Transports in *Penicillium cyclopium* W., D. Sc., thesis, Martin Luther University, Halle (GDR), 1981.
19. **Luckner, M.,** The integration of benzodiazepine and quinoline alkaloid formation into the developmental program of *Penicillium cyclopium,* in *Cell Differentiation in Microorganisms, Plants and Animals,* Nover, L. and Mothes, K., Eds., VEB Fischer, Jena, 1977, 538.
20. **Nover, L. and Luckner, M.,** Expression of secondary metabolism as part of the differentiation processes during the idiophase development of *Penicillium cyclopium* W., *Biochem. Physiol. Pflanzen,* 166, 293, 1974.
21. **Luckner, M.,** Alkaloid biosynthesis in *Penicillium cyclopium* — does it reflect general features of secondary metabolism?, *J. Nat. Prod.,* 43, 21, 1980.
22. **Nover, L., Lerbs, W., Müller, W., and Luckner, M.,** Channeling of exogenous phenylalanine to the sites of storage and the sites of alkaloid and protein biosynthesis in *Penicillium cyclopium, Biochim. Biophys. Acta,* 584, 270, 1979.
23. **Luckner, M., Diettrich, B., and Lerbs, W.,** Cellular compartmentation and channeling of secondary metabolism in microorganisms and higher plants, in *Progress in Phytochemistry,* Vol. 6, 1st ed., Reinhold R., Harborne, J. B., and Swain, T., Eds., Pergamon Press, Oxford, 1980, 103.
24. **Alibert, G., Ranjeva, R., and Boudet, A. M.,** Organisation subcellulaire des voies de synthèse des composés phenoliques, *Physiol. Vég.,* 15(2), 279, 1977.
25. **Ranjeva, R., Boudet, A. M., and Alibert, G.,** Isoenzymes and compartmentation in the regulation of phenolic metabolism, in *Regulation of Secondary Product and Plant Hormone Metabolism,* Proc. 12th FEBS Meet., Vol. 55, Luckner, M. and Schreiber, K., Eds., Pergamon Press, Oxford, 1979, 91.

25a. **Czichi, U. and Kindl, H.,** Phenylalanine ammonia lyase and cinnamic acid hydroxylases as assembled consecutive enzymes on microsomal membranes of cucumber cotyledons: cooperation and subcellular distribution, *Planta,* 134, 133, 1977.

26. **Ranjeva, R., Alibert, G., and Boudet, A. M.,** Metabolisme des composés phénoliques chez le *Petunia.* VI. Intervention des chloroplastes dans la biosynthèse de la naringénine et de l'acide chlorogènique, *Plant Sci. Lett.*, 10, 235, 1977.
27. **Pickett-Heaps, J. D.,** Xylem wall deposition. Radioautographic investigations using lignin precursors, *Protoplasma*, 65, 181, 1968.
28. **Gross, G. C., Janse, C., and Elsner, E. F.,** Involvement of malate, monophenols, and the superoxide radical in hydrogen peroxide formation by isolated cell walls from horseradish (*Armoracia lapthifolia* Gibb.), *Planta*, 136, 271, 1977.
29. **Grisebach, H.,** Lignins, in *The Biochemistry of Plants*, Vol. 7, Conn, E. E., Ed., Academic Press, New York, 1981, 457.
30. **Stafford, H. A.,** The metabolism of aromatic compounds, *Annu. Rev. Plant Physiol.*, 25, 459, 1974.
30a. **Mayer, A. M. and Harel, E.,** Polyphenol oxidases in plants, *Phytochemistry*, 18, 193, 1979.
31. **Mäder, M.,** Die Lokalisation der Peroxidase-Isoenzymgruppe G_1 in der Zellwand von Tabak-Geweben, *Planta*, 131, 11, 1976.
32. **Mäder, M., Meyer, Y., and Bopp, M.,** Zellwandregeneration und Peroxidase-Isoenzym-Synthese isolierter Protoplasten von *Nicotiana tabacum* L., *Planta*, 129, 33, 1976.
33. **Smith, D. C. C.,** Ester groups in lignin, *Nature, (London)*, 176, 267, 1955.
34. **Higuchi, T., Ito, Y., Shimada, M., and Kawamura, J.,** Chemical properties of milled wood lignin of grasses, *Phytochemistry*, 6, 1551, 1967.
35. **Gestetner, B. and Conn, E. E.,** The 2-hydroxylation of transcinnamic acid by chloroplasts of *Melilotus alba*, *Arch. Biochem. Biophys.*, 163, 617, 1974.
36. **Ranjeva, R., Alibert, G., and Boudet, A. M.,** Metabolisme des composés phénoliques des tissues de *Petunia.* V. Utilization de la phenylalanine par des chloroplastes isolés, *Plant Sci. Lett.*, 10, 225, 1977.
37. **Dhillon, D. S. and Brown, A. S.,** Localization, purification and characterization of dimethylallylpyrophosphate: umbelliferone: dimethylallyl-transferase from *Ruta graveolens*, *Arch. Biochem. Biophys.*, 177, 74, 1976.
38. **Czichi, U. and Kindl, H.,** A model of closely assembled consecutive enzymes on membranes: formation of hydroxycinnamic acids from L-phenylalanine on thylakoids of *Dunaliella marina*, *Hoppe-Seyler's Z. Physiol. Chem.*, 356, 475, 1975.
39. **Löffelhardt, W. and Kindl, H.,** The conversion of L-phenylalanine into benzoic acid on the thylakoid membrane of higher plants, *Hoppe-Seyler's Z. Physiol. Chem.*, 365, 487, 1975.
40. **Kindl, H.,** Compartments and microcompartments channelling intermediates in phenylpropanoid metabolism, in *Regulation of Secondary Product and Plant Hormone Metabolism*, Proc. 12th FEBS Meet., Vol. 55, Luckner, M. and Schreiber, K., Eds., Pergamon Press, Oxford, 1979, 49.
41. **Hagel, P. and Kindl, H.,** *p*-Hydroxybenzoate synthase: a complex associated with mitochondrial membranes of roots of *Cucumis sativus*, *FEBS Lett.*, 59, 120, 1975.
42. **Löffelhardt, W. and Kindl, H.,** Formation of benzoic acid and *p*-hydroxybenzoic acid in the blue green alga *Anacystis nidulans:* a thylakoid-bound enzyme complex analogous of the chloroplast system, *Z. Naturforsch.*, 31c, 693, 1976.
43. **Ruis, H. and Kindl, H.,** Formation of α, β-unsaturated carboxylic acids from amino acids in plant peroxisomes, *Phytochemistry*, 10, 2627, 1971.
44. **Boudet, A., Ranjeva, R., and Gadal, P.,** Propriétés allosteriques specifiques des deux isoenzymes de la phenylalanine ammoniaque-lyase chez *Quercus pedunculata*, *Phytochemistry*, 10, 997, 1971.
45. **Alibert, G., Ranjeva, R., and Boudet, A.,** Recherches sur les enzymes catalysant la formation des acides phenoliques chez *Quercus pedunculata* (Ehrh.). II. Localization intracellulaire de la phenylalanine ammoniaque-lyase, de la cinnamate 4-hydroxylase et de la "benzoate synthease", *Biochim. Biophys. Acta*, 279, 282, 1972.
46. **Charriére-Ladreix, Y.,** Sur la presence de deux isoenzymes de la phenylalanine ammonia-lyase chez *Aesculus hippocastanum*, *Phytochemistry*, 14, 1727, 1975.
47. **Quail, P. H.,** Plant cell fractionation, *Annu. Rev. Plant Physiol.*, 30, 425, 1979.
47a. **Hall, I. L. and Morre, A. L., Eds.,** *Isolation of Membranes and Organelles from Plant Cells*, Academic Press, London, 1983.
48. **Hill, A. Ch. and Rhodes, M. J. C.,** The properties of cinnamic acid-4-hydroxylase of aged sweet root disks, *Phytochemistry*, 14, 2387, 1975.
49. **Charriére-Ladreix, Y.,** *O*-methyltransferases endoplamiques des *Populus nigra*, *Phytochemistry*, 18, 43, 1979.
50. **Wilson, S. and Luckner, M.,** Cyclopenase, ein Lipoproteid der Plasmamembran von Konidiosporen des Pilzes *Penicillium cyclopium* WESTLING, *Z. Allg. Mikrobiol.*, 15, 45, 1975.
51. **Gennis, R. B., Sinensky, M., and Strominger, I. L.,** Activation of C_{55}-isoprenoid alcohol phosphokinase from *Staphylococcus aureus.* II. Biophysical studies, *J. Biol. Chem.*, 251, 1270, 1976.
52. **Sandermann, H.,** The reactivation of C_{55}-isoprenoid-alcohol-phosphokinase apoprotein by lipids. Evidence for lipid hydration in lipoprotein function, *Eur. J. Biochem.*, 43, 415, 1974.

53. **Gennis, R. B. and Strominger, I. L.,** Activation of C_{55}-isoprenoid alcohol phosphokinase from *Staphylococcus aureus*. I. Activation by phospholipids and fatty acids, *J. Biol. Chem.*, 251, 1264, 1976.
54. **Graebe, J. E., Hedden, P., Gaskin, Pl., and MacMillan, J.,** Biosynthesis of gibberellins A_{12}, A_{15}, A_{24}, A_{36} and A_{37} by a cell-free system from *Cucurbita maxima*, *Phytochemistry*, 13, 1433, 1974.
55. **Hedden, P., MacMillan, J., and O'Phinney, G.,** The metabolism of the gibberellins, *Annu. Rev. Plant Physiol.*, 29, 149, 1978.
56. **Roos, W.,** Die Stoffwechselabhängigkeit von Permeabilitätsbarrieren im Alkaloidstoffwechsel von *Penicillium cyclopium*, *Pharmazie*, 29, 78, 1974.
57. **Roos, W., Fürst, W., and Luckner, M.,** Significance of membrane permeability for synthesis and distribution of alkaloids in cultures of *Penicillium cyclopium* WESTLING, *Nova Acta Leopoldina*, 7(Suppl.), 175, 1976.
58. **Roos, W. and Luckner, M.,** ATP-dependent permeability of membrane barriers in *Penicillium cyclopium* WESTLING, *Biochem. Physiol. Pflanzen*, 171, 127, 1977.
59. **Wellburn, A. R. and Hampp, R.,** Changes in plastid and mitochondrial envelopes during greening, *Biochem. Soc. Trans.*, 5, 91, 1977.
60. **Wellburn, A. R. and Hampp, R.,** Fluxes of gibberellic and abscisic acids, together with that of adenosine-3′,5′-cyclo phosphate, across plastid envelopes during development, *Planta*, 131, 95, 1976.
61. **Wellburn, A. R. and Hampp, R.,** Uptake of mevalonate and acetate during plastid development, *Biochem. J.*, 158, 231, 1976.
62. **Wellburn, A. R. and Hampp, R.,** Movement of labelled metabolites from mitochondria to plastids during development, *Planta*, 131, 17, 1976.
63. **Hampp, R. and Wellburn, A. R.,** Changes in the permeability of the inner mitochondrial membrane associated with plastid development, *Planta*, 131, 21, 1976.
64. **Lehner, K. and Heldt, H. W.,** Dicarboxylate transport across the inner membrane of the chloroplast envelope, *Biochim. Biophys. Acta*, 501, 531, 1978.
65. **Gimmler, H., Schäfer, G., Kraminer, H., and Heber, U.,** Amino acid permeability of the chloroplast envelope as measured by light scattering, volumetry and amino acid uptake, *Planta*, 120, 47, 1974.
66. **Cobb, A. H. and Wellburn, A. R.,** Changes in plastid envelope polypeptides during chloroplast development, *Planta*, 121, 273, 1974.
67. **Cobb, A. H. and Wellburn, A. R.,** Polypeptide binding to plastid envelopes during chloroplast development, *Planta*, 129, 127, 1976.
68. **Wellburn, F. A. M. and Wellburn, A. R.,** Response of etioplasts *in situ* and isolated suspensions to pre-illumination with various combinations of red, far-red and blue light, *New Phytol.*, 72, 55, 1973.
69. **Kende, H. and Hanson, A. D.,** Relationship between ethylene evolution and senescence in morning-glory flower tissue, *Plant Physiol.*, 57, 523, 1976.
70. **Hanson, A. D. and Kende, H.,** Methionine metabolism and ethylene biosynthesis in senescent flower tissue of morning glory, *Plant Physiol.*, 57, 528, 1976.
71. **Imaseki, H. and Watanabe, A.,** Inhibition of ethylene production by osmotic shock. Further evidence for membrane control of ethylene production, *Plant Cell Physiol.*, 19, 345, 1978.
72. **Kende, H. and Hanson, A. D.,** On the role of ethylene in aging, in *Plant Growth Regulation*, Pilet, P. E., Ed., Springer-Verlag, Berlin, 1977, 172.
73. **Hahlbrock, K.., Knobloch, K. H., Kreuzaler, F., Potts, I. R. M., and Wellmann, E.,** Coordinated induction and subsequent activity changes of two groups of metabolically interrelated enzymes. Light induced synthesis of flavonoid glycosides in cell suspension cultures of *Petroselinum hortense*, *Eur. J. Biochem.*, 63, 137, 1976.
74. **Schröder, J., Kreuzaler, F., Schäfer, E., and Hahlbrock, K.,** Concomitant induction of phenylalanine ammonia-lyase and flavanone synthase mRNAs in irradiated plant cells, *J. Biol. Chem.*, 254, 57, 1979.
75. **Hahlbrock, K. and Grisebach, H.,** Enzymic controls in the biosynthesis of lignins and flavonoids, *Annu. Rev. Plant Physiol.*, 30, 105, 1979.
76. **Marmé, D.,** Phytochrome: membranes as possible sites of primary action, *Annu. Rev. Plant Physiol.*, 28, 173, 1977.
77. **Dumortier, F. and Vendrig, J.,** Effect of "penetrant-carriers" in relation to the photocontrol of anthocyanin synthesis in mung beans (*Phaseolus aureus* Ros. B.), *Z. Pflanzenphysiol.*, 87, 313, 1978.
78. **Pecket, R. C. and Hathout Bassim, T. A.,** Mechanism of phytochrome action in the control of biosynthesis of anthocyanins, *Phytochemistry*, 13, 815, 1974.
79. **Hathout Bassim, T. A. and Pecket, R. C.,** The effect of membrane stabilizers on phytochrome-controlled anthocyanin biosynthesis in *Brassica oleracea*, *Phytochemistry*, 14, 731, 1975.
80. **Cooke, R. J., Saunders, P. F., and Kendrick, R. E.,** Red-light induced production of gibberellin-like substances in homogenates of etiolated wheat leaves and in suspenions of intact etioplasts, *Planta*, 124, 319, 1976.
81. **Cooke, R. J. and Kendrick, R. E.,** Phytochrome controlled gibberellin metabolism in etioplast envelopes, *Planta*, 131, 303, 1976.

82. **Ranjeva, R., Boudet, A. M., and Faggion, R.,** Phenolic metabolism in *Petunia tissues.* IV. Properties of *p*-coumarate: CoA ligase isoenzymes, *Biochimie,* 58, 1255, 1976.
83. **Knobloch, K. H. and Hahlbrock, K.,** Isoenzymes of *p*-coumarate: CoA ligase from cell suspension cultures of *Glycine max, Eur. J. Biochem.,* 52, 311, 1975.
84. **Mackenzie, J. M., Jr., Coleman, R. A., Briggs, W. R., and Pratt, L. H.,** Reversible redistribution of phytochrome within the cell upon conversion to its physiologically active form, *Proc. Natl. Acad. Sci. U.S.A.,* 73, 799, 1975.
85. **Stafford, H.,** Compartmentation in natural product biosynthesis by multienzyme complexes, in *The Biochemistry of Plants,* Vol. 7, Conn, E. E., Ed., Academic Press, New York, 1981, 117.
86. **Conn, E. E., McFarlane, I. J., Møller, B. L., and Shimada, M.,** Channeling of intermediates during the biosynthesis of cyanogenic glycosides, in *Regulation of Secondary Product and Plant Hormone Metabolism,* Proc. 12th FEBS Meet., Vol. 55, Luckner, M. and Schreiber, K., Eds., Pergamon Press, Oxford, 1979, 63.
87. **Welch, G. R. and Gaertner, F. H.,** Influence of an aggregated multienzyme system on transient time. Kinetic evidence for substrate compartmentation by the aromatic amino acid biosynthetic complex of *Neurospora crassa, Proc. Natl. Acad. Sci. U.S.A.,* 72, 4218, 1975.
88. **Welch, G. R. and Gaertner, H. F.,** Coordinate activation of a multienzyme complex by the first substrate. Evidence for a novel regulatory mechanism in the poly-aromatic pathway of *Neurospora crassa, Arch. Biochem. Biophys.,* 172, 476, 1976.
89. **Srere, P. J. and Mosbach, K.,** Metabolic compartmentation: symbiotic, organellar, multienzymic, and microenvironmental, *Annu. Rev. Microbiol.,* 28, 61, 1974.
90. **Mosbach, K. and Mattiasson, B.,** Multistep enzyme systems, in *Methods in Enzymology,* Vol. 44, Mosbach, K., Ed., Academic Press, New York, 1976, 453.
91. **Mowbray, J. and Moses, V.,** The tentative identification in *E. coli* of a multienzyme complex with glycolytic activity, *Eur. J. Biochem.,* 66, 25, 1976.
92. **Srere, P. A., Halper, L. A., and Finkelstein, M. B.,** Interaction of citrate synthase and malate dehydrogenase, in *Microenvironments and Metabolic Compartmentation,* Srere, P. A. and Estabrook, R. W., Eds., Academic Press, New York, 1978, 419.
93. **Lynen, F.,** Structure and function of fatty acid synthetases, in *Cell Compartmentation and Metabolic Channeling,* Nover, L., Lynen, F., and Mothes, K., Eds., VEB Fischer, Jena, 1980, 125.
94. **Scallen, T. J., Schuster, M., and Dhar, A. K.,** Evidence for a noncatalytic carrier protein in cholesterol biosynthesis, *J. Biol. Chem.,* 246, 224, 1971.
95. **Vahonny, G. V., Chanderbhan, R., Noland, B. J., Irwin, D., Dennis, P., Lambeth, J. O., and Scallen, T. J.,** Sterol carrier protein$_2$, *J. Biol. Chem.,* 258, 11731, 1983.
96. **Moore, T. C., Barlow, S. A., and Coolbaugh, R. C.,** Participation of a noncatalytic "carrier" protein in the metabolism of kaurene in cell free extracts of pea seeds, *Phytochemistry,* 11, 3225, 1972.
97. **Dimroth, P., Ringelmann, E., and Lynen, F.,** 6-Methylsalicylic acid synthetase from *Penicillium patulum, Eur. J. Biochem.,* 68, 591, 1976.
98. **Light, R. J. and Hager, P. L.,** Molecular parameters of 6-methylsalicylic acid synthetase from gel filtration and sucrose density gradient centrifugation, *Arch. Biochem. Biophys.,* 125, 326, 1968.
99. **Bisswanger, H. and Schmincke-Ott, E., Eds.,** *Multifunctional Proteins,* John Wiley & Sons, New York, 1979, 217.
100. **Schweizer, E., Kniep, B., Castorph, H., and Holzner, U.,** Pantetheine-free mutants of the yeast fatty acid synthetase complex, *Eur. J. Biochem.,* 39, 352, 1973.
101. **Kleinkauf, H. and von Döhren, H.,** Nucleic acid independent synthesis of peptides, *Curr. Top. Microbiol. Immunol.,* 91, 129, 1981.
102. **Pass, L., Zimmer, T.-L., and Laland, S. G.,** On the use of affinity chromatography in demonstrating the transfer of thioester-bound D-phenylalanine from the light enzyme of gramicidine S synthetase to an acceptor site for this amino acid on the heavy enzyme, *Eur. J. Biochem.,* 47, 607, 1974.
103. **Lee, S. G. and Lipman, F.,** Isolation of amino acid activating subunit pantetheine protein complexes: their role in chain elongation in tyrocidine biosynthesis, *Proc. Natl. Acad. Sci. U.S.A.,* 74, 2343, 1977.
104. **Kleinkauf, H. and Koischwitz, H.,** Gramicidine S-synthetase, in *Multifunctional Proteins,* Bisswanger, H. and Schmincke-Ott, E., Eds., John Wiley & Sons, New York, 1980, 217.
105. **Hahlbrock, K.,** Flavonoids, in *The Biochemistry of Plants,* Vol. 7, Conn, E. E., Ed., Academic Press, New York, 1981, 425 and 433 to 437.
106. **Lee, S. G. and Lipman, F.,** Tyrocidine synthetase system, *Methods in Enzymology,* Vol. 43, Academic Press, New York, 1975, 585.
107. **McFarlane, I. J., Lees, E. M., and Conn, E. E.,** The in vitro biosynthesis of dhurrin, the cyanogenic glycoside in *Sorghum bicolor, J. Biol. Chem.,* 350, 4708, 1975.
108. **Shimada, M. and Conn, E. E.,** The enzymatic conversion of *p*-hydroxyphenylacetaldoxime to *p*-hydroxyphenylacetonitrile, *Arch. Biochem. Biophys.,* 180, 199, 1977.

109. **Mueller, W. C. and Greenwood, A. D.,** The ultrastructure of phenolics-storing cells fixed with caffein, *J. Exp. Bot.*, 29, 757, 1978.
110. **Baur, P. S. and Walkinshaw, C. H.,** Fine structure of tannin accumulations in callus cultures of *Pinus elliotti, Can. J. Bot.*, 52, 615, 1974.
111. **Barz, W. and Hösel, W.,** in *The Flavonoids,* Harborne, J. B., Mabry, T. J., and Mabry, H., Eds. Chapman & Hall, London, 1975, 916.
112. **Lang, E., Amelunxen, F., Friedrich, H., and Hörster, H.,** Morphometrische und Chemische Charakterisierung der Bildungsund Metabolisierungskinetik von "Lamiaceengerbstoffen" in *Ocimum basilicum-*Zellsuspensionskulturen, *Eur. J. Cell. Biol.*, 19, 67, 1979.
113. **Ebel, J., Ayers, A. R., and Albersheim, P.,** Host-pathogen interactions. XII. Response of suspension-cultured soybean cells to the elicitor isolated from *Phytophthora megasperma* var. *sojae,* a fungal pathogen of soybean, *Plant Physiol.*, 57, 775, 1976.
114. **Boydston, R., Paxton, J. O., and Koeppe, D. E.,** Glyceollin: a site-specific inhibitor of electron transport in isolated soybean mitochondria, *Plant Physiol.*, 72, 151, 1983.
115. **Ravanel, P., Tissut, R., and Douce, R.,** Effects of kaempferol on the oxidative properties of intact plant mitochondria, *Plant Physiol.*, 69, 375, 1982.
116. **Matile, Ph., Jans, B., and Rickenbacher, R.,** Vacuoles of *Chelidonium latex:* lysosomal property and accumulation of alkaloids, *Biochem. Physiol. Pflanzen,* 161, 447, 1970.
117. **Matile, Ph.,** Localization of alkaloids and mechanism of their accumulation in vacuoles of *Chelidonium majus laticifers,* in *Secondary Metabolism and Coevolution,* Nova Acta Leopoldina, Suppl. 7, 139, 1976.
118. **Shiraishi, T., Oku, H., Isono, M., and Ouchi, S.,** The injurious effect of pisatin on the plasma membrane of pea, *Plant Cell Physiol.*, 16, 939, 1975.
119. **Lyon, G. D.,** Evidence that the toxic effect of rishitin may be due to membrane damage, *J. Exp. Bot.*, 31, 957, 1980.
120. **Hargreaves, J. A.,** A possible mechanism for the phytotoxicity of the phytoalexin phaseollin, *Physiol. Plant Pathol.*, 16, 351, 1980.
121. **Oba, K., Conn, E. E., Canut, H., and Boudet, A. M.,** Subcellular localization of 2-(β-D-glucosyloxy)-cinnamic acids and the related β-glucosidase in leaves of *Melilotus alba* Dess., *Plant Physiol.*, 68, 1359, 1981.
122. **Lüthy, B. and Matile, Ph.,** The mustard oil bomb: rectified analysis of the subcellular organization of the myrosinase system, *Biochem. Physiol. Pflanzen,* 179, 5, 1984.
123. **Saunders, J. A. and Conn, E. E.,** Presence of the cyanogenic glucoside dhurrin in isolated vacuoles from *Sorghum, Plant Physiol.*, 61, 154, 1978.
124. **Kojima, M., Poulton, J. E., Thayer, S., and Conn, E. E.,** Tissue distribution of dhurrin and enzymes involved in its metabolism in leaves of *Sorghum bicolor, Plant Physiol.*, 63, 1022, 1979.
125. **Thayer, S. S. and Conn, E. E.,** Subcellular localization of dhurrin-β-glucosidase and hydroxynitrile lyase in the mesophyll cells of *Sorghum* leaf blades, *Plant Physiol.*, 67, 617, 1981.
126. **Mitzka-Schnabel, U. and Rau, W.,** The subcellular distribution of carotenoids in *Neurospora crassa, Phytochemistry,* 19, 1409, 1980.
127. **Schnepf, E.,** Morphology and cytology of storage spaces, *Nova Acta Leopoldina,* 7,(Suppl.), 23, 1979.
128. **Schulze, C., Schnepf, E., and Mothes, K.,** Über die Lokalisation der Kautschukpartikel in Verschiedenen Typen von Milchröhren, *Flora,* A158, 458, 1967.
129. **Neumann, D., Lösecke, W., Maier, W., and Gröger, D.,** Localization of alkaloids in sclerotia and suspension cultures of *Claviceps purpurea, Biochem. Physiol. Pflanzen,* 174, 504, 1979.
130. **Costes, C., Burghoffer, C., Carrayol, E., Ducet, C., and Diano, M.,** Occurrence of carotinoids in non-plastidial materials from potato tuber cells, *Plant Sci. Lett.*, 6, 253, 1976.
130a. **Pecket, R. C. and Small, C. J.,** Occurrence, location and development of anthocyanoplasts, *Phytochemistry,* 19, 2571, 1980.
130b. **Small, C. J. and Pecket, R. C.,** The ultrastructure of anthocyanoplasts in red cabbage, *Planta,* 154, 97, 1982.
130c. **Neumann, D.,** Subcellular localization of anthocyanins in red cabbage seedlings, *Biochem. Physiol. Pflanzen,* 178, 405, 1983.
131. **Rebeiz, C. A. and Castelfranco, P. A.,** Protochlorophyll and chlorophyll biosynthesis in cell free systems from higher plants, *Annu. Rev. Plant Physiol.*, 24, 129, 1973.
132. **Browning, G. and Saunders, P. F.,** Membrane localized gibberellins A9 and A4 in wheat chloroplasts, *Nature, (London),* 265, 375, 1977.
133. **Matile, Ph.,** Das toxische Kompartiment der Pflanzenzelle, *Naturwissenschaften,* 71, 18, 1984.
134. **Matile, Ph. and Wiemken, A.,** Interactions between cytoplasm and vacuole, in *Encyclopedia of Plant Physiology,* (New Series), Vol. 3, Stocking, C. R. and Heber, U., Eds., Springer-Verlag, Berlin, 1976, 255.
135. **Northcote, D. H.,** The synthesis and metabolic control of polysaccharides and lignin during differentiation of plant cells, *Essays Biochem.*, 5, 89, 1969.

136. **Delmer, D. P.,** Biosynthesis of cellulose, *Adv. Carbohydr. Chem. Biochem.*, 41, 105, 1983.
137. **Brown, R. M., Franke, W. W., Kleinig, H., Falk, H., and Sitte, P.,** Scale formation in chrysophycean algae. I. Cellulosic and noncellulosic wall components made by the Golgi apparatus, *J. Cell. Biol.*, 45, 246, 1970.
138. **Herth, W., Kuppel, A., Franke, W. W., and Brown, R. M.,** The ultrastructure of the scale cellulose from *Pleurochrysis scherffelii* under various experimental conditions, *Cytobiologie*, 10, 268, 1975.
139. **Mollenhauer, H. H. and Morré, D. J.,** Golgi apparatus and plant secretion, *Annu. Rev. Plant Physiol.*, 17, 27, 1966.
140. **Ray, P. M., Eisinger, W. R., and Robinson, D. G.,** Organelles involved in cell wall polysaccharide formation and transport in pea cells, *Ber. Deutsch. Bot. Ges.*, 89, 121, 1976.
141. **Wright, K. and Northcote, D. H.,** Identification of β,1-4-glucan chains as a part of a fraction of slime synthesized within the dictyosomes of maize root caps, *Protoplasms*, 88, 225, 1976.
142. **Schnepf, E.,** Sezernierende und Exzernierende Zellen bei Pflanzen, in *Grundlagen der Cytologie*, Hirsch, C. C., Ruska, H., and Sitte, P., Eds., VEB Fischer, Jena, 1973, 461.
143. **Willemse, M. T. M.,** Morphological and fluorescence microscopical investigation on sporopollenin formation at *Pinus sylvestris* and *Gasteria verrucosa*, in *Sporopollenin*, Brooks, J., et al., Eds., Academic Press, London, 1971, 68.
144. **Charriére-Ladreix, Y.,** Intracellular distribution of flavonoids in glandular cells, in *Regulation of Secondary Product and Plant Hormone Metabolism*, Proc. 12th FEBS Meet., Vol. 55, Luckner, M. and Schreiber, K., Eds., Pergamon Press, Oxford, 1979, 101.
145. **Manheim, B. S. and Mulroy, T. W.,** Triterpenoids in epicuticular waxes of *Dudleya* species, *Phytochemistry*, 17, 1799, 1978.
146. **Schnepf, E.,** Die Kompartimentierung der Zelle in morphologischer Sicht, *Biol. Rundschau*, 4, 259, 1966.
147. **Kurylo-Borowska, Z.,** Localization of edeine synthetase within *Bacillus brevis* Vm 4, *Biochim. Biophys. Acta*, 399, 31, 1975.
148. **Robinson, D. G.,** Plant cell wall synthesis, *Adv., Bot. Res.*, 5, 89, 1977.
149. **Schnepf, E., Röderer, G., and Herth, W.,** The formation of the fibrils in the lorica of *Poteriochromonas stipitata*, *Planta*, 125, 45, 1975.
150. **Kiermayer, O.,** Biomembranen als Träger morphogenetischer Information, Untersuchungen bei der Grünalge Micrasterias, *Naturwiss. Rundsch.*, 30, 161, 1977.
151. **Robinson, D. G., Eisinger, W. R., and Ray, P. M.,** Dynamics of the Golgi system in wall matrix polysaccharide synthesis and secretion by pea cells, *Ber. Deutsch. Bot. Ges.*, 89, 147, 1976.
152. **Chafe, S. C. and Durzan, D. J.,** Tannin inclusions in cell suspension cultures of white spruce, *Planta*, 113, 251, 1973.
153. **Morré, D. J., Kartenbeck, J., and Franke, W. W.,** Membrane flow and interconversions among endomembranes, *Biochim. Biophys. Acta*, 559, 72, 1979.
154. **Morré, D. J. and Mollenhauer, H. H.,** Interactions among cytoplasm, endomembranes and the cell surface, in *Encyclopedia of Plant Physiology*, (New Series), Vol. 3, Stocking, C. R. and Heber, U., Eds., Springer-Verlag, Berlin, 1976, 288.
155. **Saunders, J. A.,** Investigation of vacuoles isolated from tobacco. I. Quantitation of nicotine, *Plant Physiol.*, 64, 74, 1979.
156. **Kurkdijan, A.,** Absorption and accumulation of nicotine by *Acer pseudoplatanus* and *Nicotiana tabacum* cells, *Physiol. Vég.*, 20, 73, 1982.
157. **Neumann, D., Krauß, G., Hieke, M., and Gröger, D.,** Indole alkaloid formation and storage in cell suspension cultures of *Catharanthus roseus*, *Plant Medica*, 48, 20, 1983.
158. **Müller, E., Nelles, A., and Neumann, D.,** Beiträge zur Physiologie der Alkaloide. I. Aufnahme von Nikotin in Gewebe von *Nicotiana rustica* L., *Biochem. Physiol. Pflanzen*, 162, 272, 1971.
159. **Brown, S., Renaudin, J. P., Prevot, Ch., and Guern, J.,** Flow cytometry and sorting of plant protoplasts: technical problems and physiological results from a study of pH and alkaloids in *Catharanthus roseus*, *Physiol. Vég.*, 22, 1984. in press.
160. **Müller, E., Neumann, D., and Nelles, A.,** Alkaloid transport and distribution in *Nicotiana* and *Macleaya*. Diversity and unity, in *Secondary Metabolism and Coevolution*, Nova Acta Leopoldina, Suppl. 7, 133, 1976.
161. **Wagner, G. J. and Liu, W.,** An active proton pump of intact vacuoles isolated from *Tulipa petals*, *Biochim. Biophys. Acta*, 689, 261, 1982.
162. **Marin, B. and Blasco, F.,** Further evidence for the proton pumping work of tonoplast ATPase from *Hevea latex* vacuome, *Biochem. Biophys. Res. Commun.*, 105, 354, 1982.
163. **Kakinuma, Y., Ohsumi, Y., and Auraku, Y.,** Properties of H^+-translocating ATPase in vacuolar membranes of *Saccharomyces cerevisiae*, *J. Biol. Chem.*, 256, 10,859, 1981.
164. **Smith, J. A. C., Uribe, E. G., Ball, E., Heuer, S., and Lüttge, U.,** Characterization of the vacuolar ATPase activity of the crassulacean-acid-metabolism plant *Kalanchoë daigremontiana*, *Eur. J. Biochem.*, 441, 415, 1984.

165. **Doll, S., Effelsberg, U., and Willenbrink, J.,** Characteristics of sucrose uptake system of vacuoles isolated from red beet tissue. Relationships between ion gradients and sucrose transport, in *Plasmalemma and Tonoplast,* Marmé, D., Marrè, R., and Hertel, R., Eds., Elsevier Biomedical, Amsterdam, 1982, 217.
166. **MacRobbie, E. A. C.,** Vacuoles: the framework, in *Plant Organelles,* Reid, E., Ed., Ellis Horwood, Chichester, 1979, 62.
167. **Wagner, G. J.,** Enzymic and protein character of tonoplast from *Hippeastrum* vacuoles, *Plant Physiol.,* 68, 499, 1981.
168. **Werner, Ch. and Matile, Ph.,** Accumulation of coumarylglycerols in vacuoles of barley mesophyll protoplasts, *J. Plant. Physiol.,* 118, 237, 1985.
169. **Deus-Neumann, B. and Zenk, M. H.,** A highly selective alkaloid uptake system in vacuoles of higher plants, *Planta,* 162, 250, 1984.

Chapter 4

AFLATOXIN BIOSYNTHESIS

T. A. V. Subramanian and R.K. Bhatnagar

TABLE OF CONTENTS

I. INTRODUCTION

Aspergilli are ubiquitous microorganisms which can grow on a wide variety of substrates. The large ecological amplitude of aspergilli confers upon them the ability to grow under wide extremes of temperature, humidity, and other environmental factors. It also gives them a distinct survival advantage over other microorganisms. It is for these reasons that aspergilli have gained tremendous importance in human life. The filamentous form of aspergilli represent an important link between well-differentiated eukaryotes and unicellular prokaryotes. With the two important evolutionary principles, namely, economy of the system and unity in diversity, operating at all times on all organisms, it is logical to presume that the metabolism of aspergilli broadly represents both eukaryotes and prokaryotes. With this in view, here we present regulatory parameters that are operative in the metabolism of aspergillus in controlling the biosynthesis of a secondary metabolite, aflatoxin.

Aspergilli have caused immense damage by causing either aspergillosis or by elaborating toxic compounds on food materials. Among several toxic compounds produced by aspergilli, aflatoxins have been a subject of thorough investigation because of their notoriety. Aflatoxins, like antibiotics, are secondary metabolites. Though genetic and metabolic regulation of several antibiotics have been worked out, not much is known about the control of aflatoxin biosynthesis. Like primary metabolites, secondary metabolites are also subjected to stringent metabolic control, such as catabolite repression, feedback control, and channelization of primary metabolites. We, in our laboratory, have been engaged in elucidating metabolic regulatory parameters governing the biosynthesis of aflatoxins. We believe that the regulatory parameters discussed here for *Aspergillus parasiticus* will broadly be the same for other microbial systems.

II. GROWTH DYNAMICS OF AFLATOXIN BIOSYNTHESIS

Aflatoxins are a group of secondary metabolites produced by several species of *Aspergillus flavus* subgroup. In the biosynthesis of aflatoxins, trace metal ions play an important role. The most important trace metal having a direct effect on aflatoxin biosynthesis is zinc. Earlier work from our laboratory demonstrated a requirement for asparagine for optimal aflatoxin production by *A. parasiticus* in a chemically defined medium. To understand the mechanism of zinc and asparagine stimulation of aflatoxin biosynthesis, studies have been carried out using a chemically defined medium, sucrose-low salts (SLS),[1] and medium devoid of zinc or asparagine. Studies of Detroy and Hesseltine[2] on aflatoxin biosynthesis led to the establishment of the following different growth periods: (1) exponential (48 hr), (2) transitional (60 hr), (3) early stationary (72 hr), (4) mid-stationary (96 hr), and (5) late stationary (120 hr). The various parameters discussed in the subsequent sections have been monitored at the above-mentioned hours.

The amount of aflatoxin produced by *A. parasiticus* both in the medium and the mycelium is shown in Figure 1A.[3] It is evident that zinc and asparagine deficiency resulted in marked reduction in the production of aflatoxins. During the mid-stationary growth phase, the total amount of aflatoxin in AD cultures (cultures grown in asparagine-deficient SLS medium) is 25% of SLS cultures, while in ZD cultures (cultures grown in zinc-deficient SLS medium) it is reduced to 10%. In all three media used, detectable amounts of aflatoxins are produced from exponential growth phase onwards (48 hr). In SLS cultures, the amount of aflatoxins in the medium is consistently higher than the amount present in mycelium. In contrast, aflatoxins are preferentially located in the mycelium under zinc and asparagine deficiency. This has been attributed to the inability of the two cultures to form hypothetical aflatoxin binding factor, which helps in the secretion of aflatoxin from the mycelium into the medium.

Under zinc and asparagine deficiency, fungal growth is also reduced proportionately to aflatoxin production and suggests that growth and aflatoxin production are interlinked. To

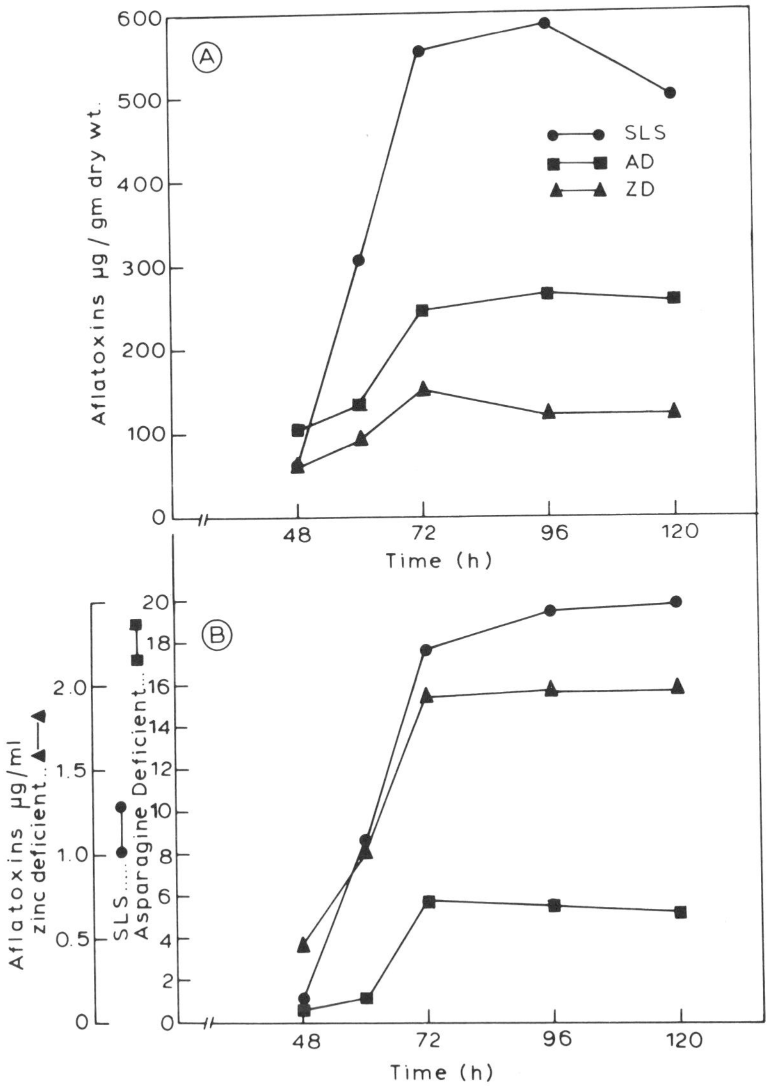

FIGURE 1. Specific aflatoxin production in *A. parasiticus*. (From Bhatnagar, R. K., Ahmadr, S., Mukerji, K. G., and Venkitasubramanian, T. A., *Appl. Environ. Microbiol.*, 44, 579, 1982. With permission.)

understand the relationship, specific aflatoxin production in mycelium (micrograms per gram dry weight) and in the medium (micrograms per milliliter) was calculated and is shown in Figure 1B.[3] It is evident that a decrease in the aflatoxin content in AD and ZD cultures is not related to the observed decrease in growth. The results clearly demonstrate that the effects of zinc and asparagine deficiency on aflatoxin biosynthesis are independent of their effects on fungal growth.

The exact mechanism by which zinc and asparagine stimulate aflatoxin production is not clearly understood at present. It is possible that asparagine serves a dual purpose. Apart from being a secondary nitrogen source, its degradation to pyruvate and finally to acetyl-CoA may lead to higher aflatoxin production. The stimulatory role of zinc in enhancing secondary metabolism is very well known, but the exact mechanism or its mode of action is not known.

From the studies on biosynthesis of secondary metabolites in relation to growth, two

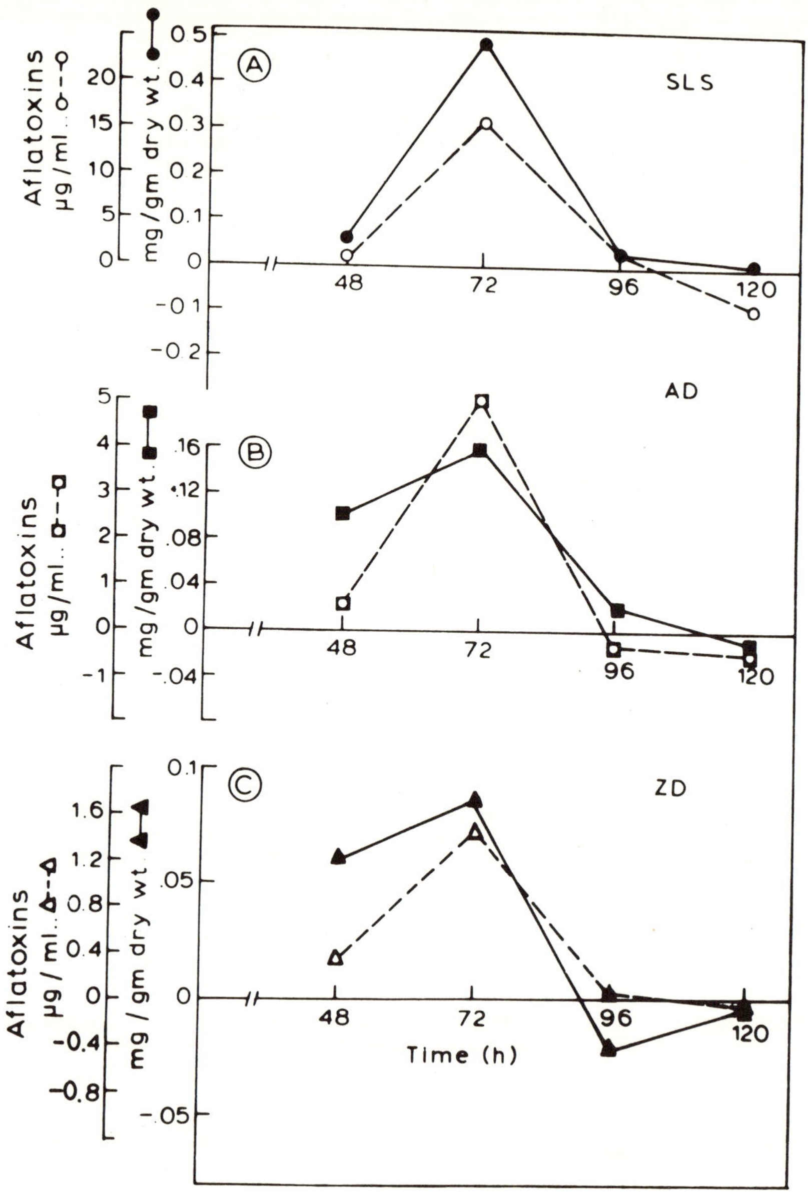

FIGURE 2. Change in specific aflatoxin production with respect to time in *A. parasiticus*. (From Bhatnagar, R. K., Ahmad, S., Mukerji, K. G., and Venkitasubramanian, T. A., *Appl. Environ. Microbiol.*, 44, 579, 1982. With permission.)

distinct patterns emerge: (1) the production of secondary metabolite begins with the exhaustion of an essential nutrient leading to the cessation of growth and (2) the production phase of the secondary metabolite is associated with the rapid growth phase.[4] To understand the relationship between aflatoxin production and fungal growth, specific aflatoxin produced in the medium and the mycelium per 24 hr of growth was calculated and is shown in Figure 2.[3] Also, increase in the dry weight per 24 hr was calculated and is shown in Figure 3. It is evident from Figure 3[5] that the maximum increase in dry weight occurred between 48 and 72 hr. Subsequently, though the total dry weight increases (Figure 3A), the increase between 72 and 96 hr is less than that between 48 and 72 hr. The net increase in dry weight

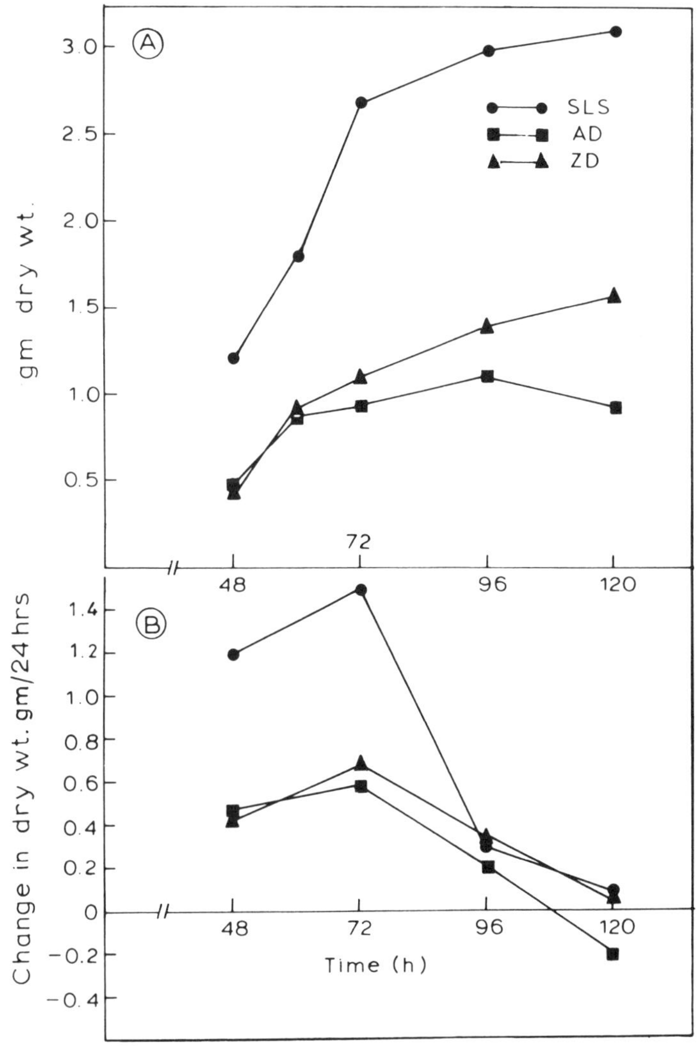

FIGURE 3. Increase in dry weight with respect to time in *A. parasiticus*. (A) The dry weight of mycelium was determined by drying a known amount of mycelium at 80 ± 1°C to a constant weight. (B) Change in dry weight per 24 hr.

between 96 and 120 hr in SLS and ZD cultures is marginal, while in AD cultures a negative value is obtained. This further substantiates that asparagine is necessary for continued fungal growth. In all three media studied, maximum specific aflatoxin production occurred at 72 hr (Figure 2). The production rate of aflatoxins decreases drastically from 72 hr onwards. The data clearly establish that active aflatoxin formation takes place along with active vegetative growth.

III. REGULATION OF AFLATOXIN BIOSYNTHESIS BY NUCLEOTIDES

The three important components that control cellular metabolism are adenosine phosphates, nicotinamide adenine dinucleotide, and nicotinamide adenine dinucleotide phosphate. Adenosine phosphates bring energy from substrate degradation and oxidative phosphorylation to anabolism, and NADH brings reduction equivalents from substrates to oxidative phosphorylation or to precursors of fermentation products. NADPH brings reducing equivalents from pentose phosphate shunt, isocitrate dehydrogenase, malic enzyme, or energy-dependent transhydrogenase to anabolism. The adenine nucleotides are intermediates in the biosynthesis of nucleic acids and nucleotide cofactors such as NAD, NADP, FAD, and CoA. These nucleotides play a stoichiometric correlation of metabolic processes and regulate the activities of a very large number of enzymes. In our laboratory, we have estimated the levels of pyridine and adenine nucleotides at different stages of growth and correlated them with aflatoxin biosynthesis. In the subsequent section we will discuss the role of these nucleotides separately.

A. Bioenergetics of Aflatoxin Biosynthesis

Regulation of microbial metabolism through the adenylic acid pool is now better understood. The levels of ATP, ADP, and AMP at varying periods of growth on different media are shown in Figures 4 and 5.[6] It is evident that in the SLS medium, as well as in the same medium deficient in zinc or asparagine, ATP levels were maxiumum at 45 hr and thereafter decreased gradually. The decrease was more marked in the transitional phase (60 hr), whereas in the stationary phase (72 to 96 hr) the levels were more or less constant. However, in the asparagine-deficient medium, the ATP level at 48 hr was not as significant as that in the other media. In the zinc-deficient medium, ATP levels at 45 and 60 hr were very similar to, but at 72 and 96 hr were much less than, the levels for corresponding periods of growth on SLS medium.

ADP levels in asparagine-deficient medium increased dramatically with growth; this increase was not so marked in the other two media (Figure 4). AMP levels, which were low in the exponential phase, increased in the transitional and stationary phases (Figure 5). Their levels were higher in the SLS medium than in the two deficient media. Earlier work from our laboratory by Gupta et al.[7] had already established that for aflatoxin biosynthesis, high levels of ATP were not required and low AMP levels were favorable. The recent experiments (Figures 4 and 5) are in agreement with the earlier reports. It is known that under zinc deficiency there is an increased lipid formation and reduced aflatoxin formation (to be discussed in detail in the following section). Low levels of AMP are reported to favor lipid synthesis,[8] and therefore would be unfavorable for aflatoxin production. On the other hand, an increase in AMP levels in both the asparagine and zinc-deficient cultures from 45 to 60 hr is observed to be much more than in the SLS medium (Figure 5). The increase in ADP levels (Figure 4) which occurs from the exponential to the stationary phase is probably due to the conversion of ATP to ADP.

It has been suggested that the adenylate energy charge is a good indicator of the metabolic state of cell.[8] Several derived parameters, like K_{app} (which is the apparent equilibrium constant for the adenylate kinase reaction), total adenylate pool, and ATP to ADP ratio (which is an index of the oxidative metabolism of the organism) have been studied (Table 1). The total adenylic acid pool (ATP + ADP + AMP) is seen to be in the range of 1.0 to 4.0 as also reported for a large number of other microorganisms.[9] In asparagine-deficient cultures, the total adenylate pool doubles from 45 to 96 hr, largely because of an increase in ADP and AMP levels. On the other hand, the pool size of zinc-deficient cultures was relatively small since aflatoxin production is maximum in SLS medium (Figure 1). It is probable that a constant pool size during the stationary phase is favorable for aflatoxin biosynthesis, and fluctuations in the total adenylate levels may be inhibitory to aflatoxin formation.

In all three media, the ATP to ADP ratio of *A. parasiticus* was higher in the exponential phase than in the stationary phase. It is very well known that secondary metabolism is nonreductive. The lower oxidative state (ATP to ADP) of the fungus during the exponential phase in the SLS medium may block acetate oxidation via the TCA cycle and NADPH oxidation, thereby resulting in the accumulation of acetate and NADPH. The resultant elevated levels of acetate and NADPH may favor the onset of aflatoxin biosynthesis. Also,

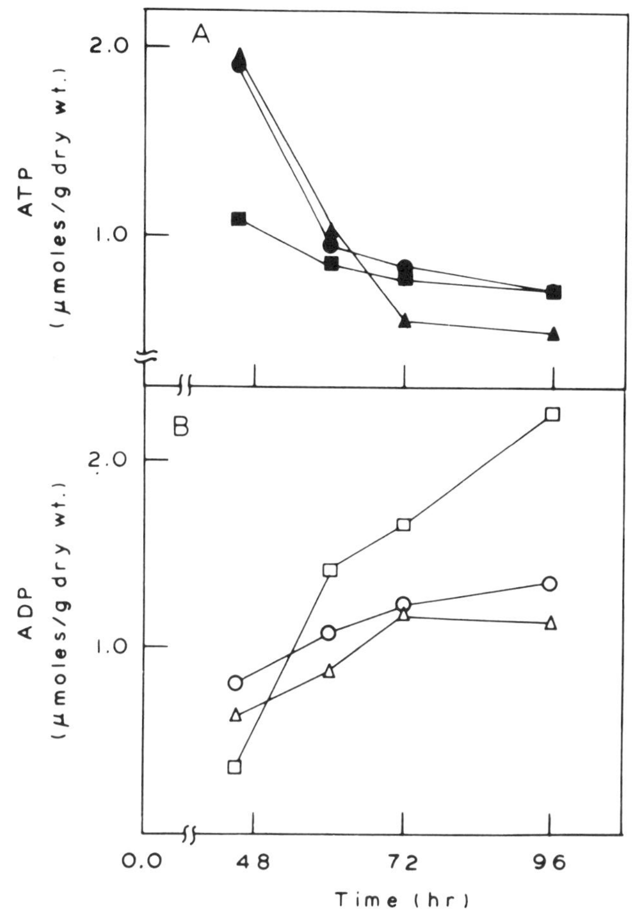

FIGURE 4. Changes in the intracellular levels of ATP and ADP in growing cells of *A. parasiticus* during aflatoxin production. (●—●) Sucrose-low salts (SLS) medium. (■—■) Asparagine-deficient SLS (AD) medium. (△—△) Zinc-deficient SLS (ZD) medium.

it is suggested that the higher oxidative state of the fungus is favorable for the maintenance of aflatoxin biosynthesis. The aparent equilibrium constant of the adenylate reaction varies, depending on the prevailing energy charge and the Mg^{2+} to total adenine nucleotide ratio. The lower values of K_{app} with growth reflect lower levels of ATP and increased ADP and AMP levels.

The correlation between energy charge, growth, and secondary metabolism is well established. Energy charge values are high during the exponential phase (0.7 to 0.9) and drop to 0.5 with the onset of secondary metabolism during secondary (growth) phase. We have also noticed a similar drop in energy charge values during the aflatoxin-producing phase in *A. parasiticus*. For typical aerobic heterotrophic eukaryotes, a high value of the adenylate energy charge facilitates carboxylation of pyruvate leading to the biosynthetic materials (amino acids) and decreases the affinity of the pyruvate dehydrogenase complex for pyruvate.[8]

Our experiments[6] have shown that the energy charge values in the SLS and asparagine-deficient media during exponential (45 hr) and transitional (60 hr) phases were lower than in zinc-deficient cultures (Figure 5). During the stationary phase, there was a drop in the energy charge values under all the cultural conditions employed. The low energy charge values observed in the stationary phase thus result in a decrease in carboxylation of pyruvate

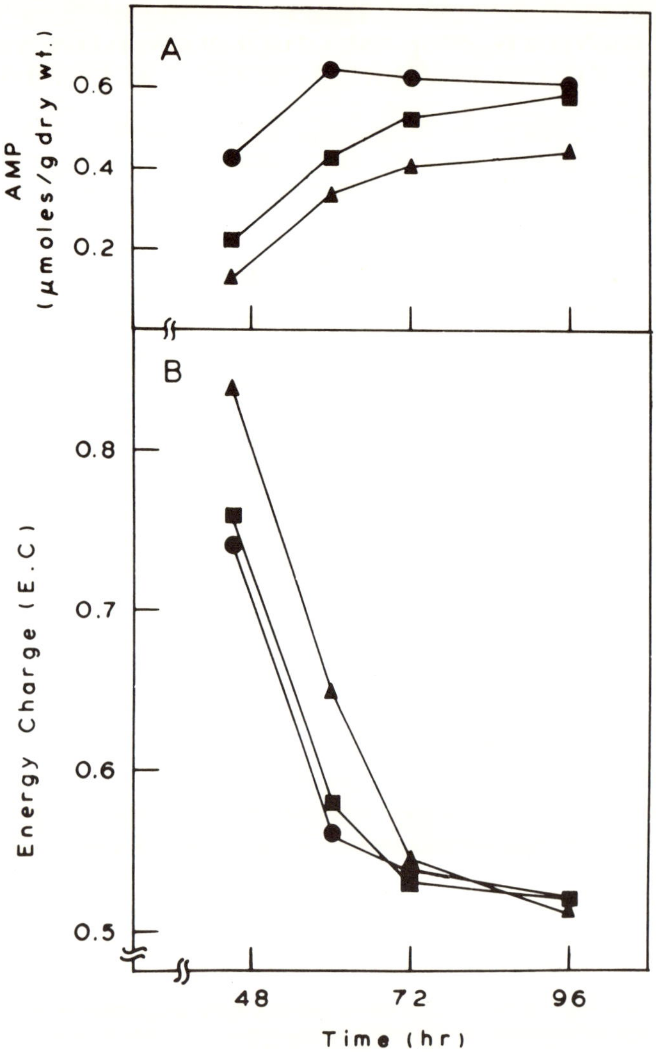

FIGURE 5. (A) Changes in the intracellular levels of AMP in growing cells of *A. parasiticus* during aflatoxin production. (B) Changes in energy charge (EC) in growing cells of *A. parasiticus* during aflatoxin production. (●—●) Sucrose-low salts (SLS) medium. (■—■) Asparagine-deficient SLS (AD) medium. (▲—▲) Zinc-deficient SLS (ZD) medium.

to oxaloacetate and an increase in the oxidative decarboxylation of pyruvate to acetyl-CoA. With the fall in the concentration of oxaloacetate, the rate of citrate synthase reaction is decreased because of substrate limitation; therefore, a decrease in the consumption of acetyl-CoA ensues. The net result is the accumulation of acetyl-CoA during the stationary phase, which is a direct precursor for aflatoxin biosynthesis. Whether ATP citrate lyase plays a significant role in the transfer of acetyl-CoA that accumulates in cytosol from mitochondria remains to be seen. However, it is unlikely because in the stationary phase citrate lyase activity is bound to decline[10] due to the fall in the ATP levels and the specific activity of citrate synthase (Figure 6).[11]

The incorporation of (8-C^{14}) adenine into different nucleotides in *A. parasiticus* under varying growth conditions is shown in Table 2.[6]

Incorporation into ATP and ADP was considerably lower in asparagine-deficient cultures

Table 1
TOTAL ADENYLATE POOL AND OTHER DERIVED FUNCTIONS IN *A. PARASITICUS* UNDER DIFFERENT CULTURE CONDITIONS

Medium used	Period of growth (hr)	Total adenine pool (ATP + ADP + AMP) (μmol/g dry wt)	ATP/ADP	$\frac{K_{app}(ATP)(AMP)}{(ADP)^2}$
SLS	45	3.15	2.36	1.25
	60	2.67	0.89	0.54
	72	2.69	0.69	0.36
	96	2.68	0.52	0.24
Asparagine deficient	45	1.67	3.03	1.85
	60	2.11	0.60	0.18
	72	2.99	0.43	0.13
	96	3.56	0.35	0.09
Zinc deficient	45	2.72	3.15	0.65
	60	2.25	1.18	0.45
	72	2.16	0.47	0.16
	96	2.07	0.43	0.17

Note: *A. parasiticus* was grown as stationary culture at 28 ± 1°C for different growth periods. A known amount of rapidly harvested mycelium was pressed between plates of aluminum foil precooled in liquid nitrogen. Adenine nucleotides were extracted from the fungal mycelium. ATP was determined by the luciferase reaction by use of scintillation counter as photometer. ADP and AMP were estimated together and the derived functions were then calculated. The values here represent the average of four separate experiments run in duplicate.

than in the SLS cultures, but the reverse was true for AMP. In general, the incorporation into ADP and AMP in the stationary phase (72 and 96 hr) was lower than that in the exponential and transitional phases (45 and 60 hr) in all three media employed; the only exception was the AMP level in zinc-deficient cultures. The higher adenine incorporation into various nucleotides signifies increased turnover of these compounds. In asparagine-deficient cultures, though the net synthesis of AMP is high, its utilization in metabolic pathways (like lipid biosynthesis) is so high that the steady state concentrations are quite low. In zinc-deficient cultures, on the other hand, the net synthesis of AMP is low and its utilization in other metabolic pathways further reduces its steady state level and for aflatoxin biosynthesis. The reverse is true in SLS cultures. The net synthesis of AMP is moderate and its steady state levels are also high, thereby showing that more AMP is available for aflatoxin biosynthesis. It is quite apparent that a high rate of formation of ATP and ADP is a prerequisite for continued aflatoxin biosynthesis during the stationary growth phase. Any decrease in the rate of ATP production as observed in the zinc- and asparagine-deficient cultures probably hampers aflatoxin biosynthesis.

B. Nicotinamide Adenine Dinucleotides and Redox State Regulation of Aflatoxin Biosynthesis

The requirement of NADPH for aflatoxin biosynthesis has been demonstrated unequivocally. Shih and Marth[12] observed increased levels of aflatoxin with an increasing amount of NADPH. They further demonstrated that sodium azide, an inhibitor of terminal electron transfer chain, resulted in the accumulaiton of NADPH and an increase in the synthesis of lipids and aflatoxins. Subsequently, conversion of an intermediate of aflatoxin biosynthesis, sterigmatocystin, to aflatoxins in a cell-free extract was shown to be NADPH dependent.[13] In our own laboratory, Tyagi et al[14] used spheroplast lysates to elucidate the role of NADPH in aflatoxin biosynthesis. A low concentration of NADPH (1 and 10 μM) failed to stimulate

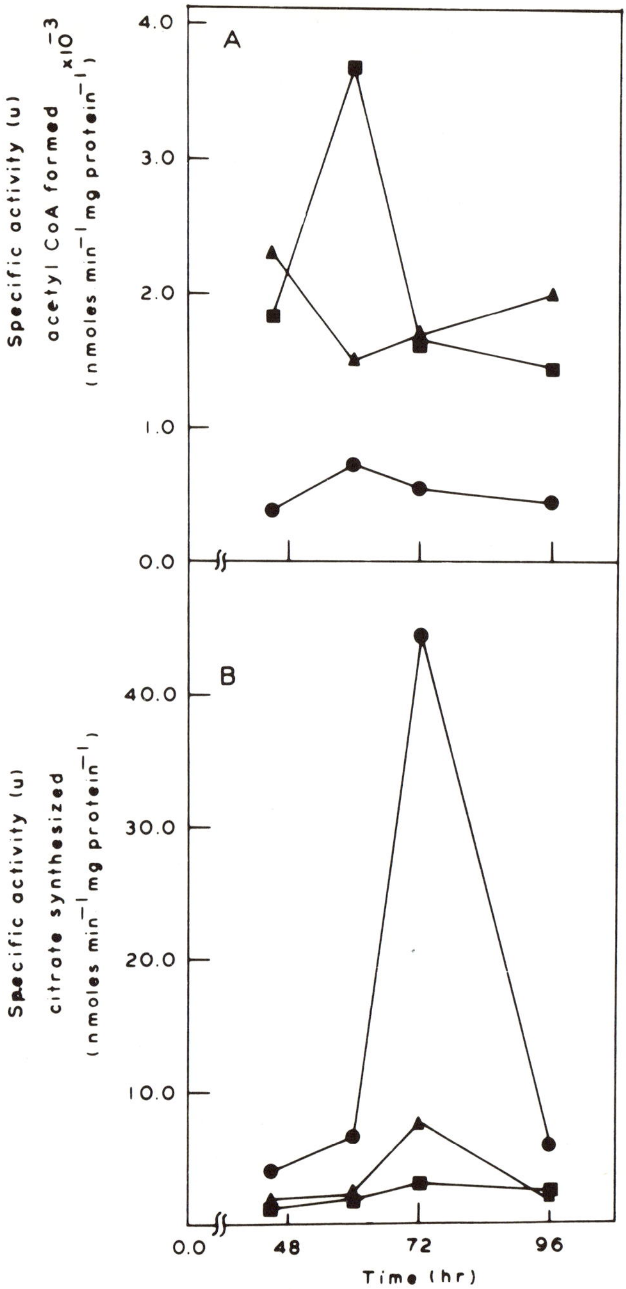

FIGURE 6. Activity of citrate synthetase in *A. parasiticus*. (●—●) Sucrose-low salts (SLS) medium. (■—■) Asparagine-deficient SLS (AD) medium. (▲—▲) Zinc-deficient SLS (ZD) medium.

acetate incorporation into aflatoxins (Table 3). On the other hand, 100 μM effected a nearly 50% stimulation of *de novo* aflatoxin synthesis from labeled acetate. It is likely that low concentrations of NADPH diverted the label to fatty acid synthesis.

Both aflatoxins and lipids are known to be acetate derived, and it has been accepted that, to a certain level, they share a common pathway. Since NADPH is used for the biosynthesis of both lipids and aflatoxins, Bhatnagar[5] investigated the relative levels of both aflatoxins

Table 2
INCORPORATION OF (8-^{14}C) ADENINE INTO VARIOUS NUCLEOTIDES

Medium used	Period of growth (hr)	Specific activity (dpm/mg adenylic acid) ATP	ADP	AMP
SLS	45	42,420	153,440	7,290
	60	47,640	172,620	5,150
	72	41,220	20,010	3,560
	96	49,980	36,180	2,810
Asparagine deficient	45	1,870	3,910	12,120
	60	1,690	2,370	10,830
	72	2,190	880	7,960
	96	2,210	1,690	6,940
Zinc deficient	45	80,850	120,610	4,050
	60	18,520	3,710	2,210
	72	1,130	3,390	5,950
	96	950	1,290	2,565

Note: 2 μCi of adenine (8-^{14}C) hydrochloride was added per 100 mℓ medium for incorporation studies. Flasks were maintained at 28 ± 1°C for different periods of growth. The adenylic acids were extracted. The ATP, ADP, and AMP were separated in a paper chromatogram using solvent system, ethanol: 0.1 *M* boric acid (75:30, V/V). The spots were viewed under ultraviolet light and the respective areas were cut and counted with 10 mℓ of toluene-based scintillation cocktail. ATP, ADP, and AMP were determined as dpm/mg adenine nucleotide. The values represented are the average of two separate experiments run in duplicate.

Table 3
EFFECT OF NADPH ON THE INCORPORATION OF ACETATE-1-^{14}C INTO AFLATOXINS BY SPHEROPLAST LYSATES

Compound added	Concentration (μ*M*)	Radioactivity incorporated into aflatoxins (counts/min/mg protein)	Change in radioactivity incorporated (%)
Nil	—	2740	—
NADPH	1	1300	−53
	10	2315	−16
	100	4075	+49

Note: Spheroplasts prepared from *A. parasiticus* NRRL 3240 grown on SLS medium on shake cultures at 28 ± 1°C for 96 hr were suspended in 0.02 *M* potassium phosphate buffers, pH 6.2. Coenzymes were added to the lysates at the specified concentrations. After incubation at 28 ± 1°C for 2 hr, the radioactivity incorporation into aflatoxin was measured. The results reported are the average of three separate experiments, the values varying by not more than 5%.

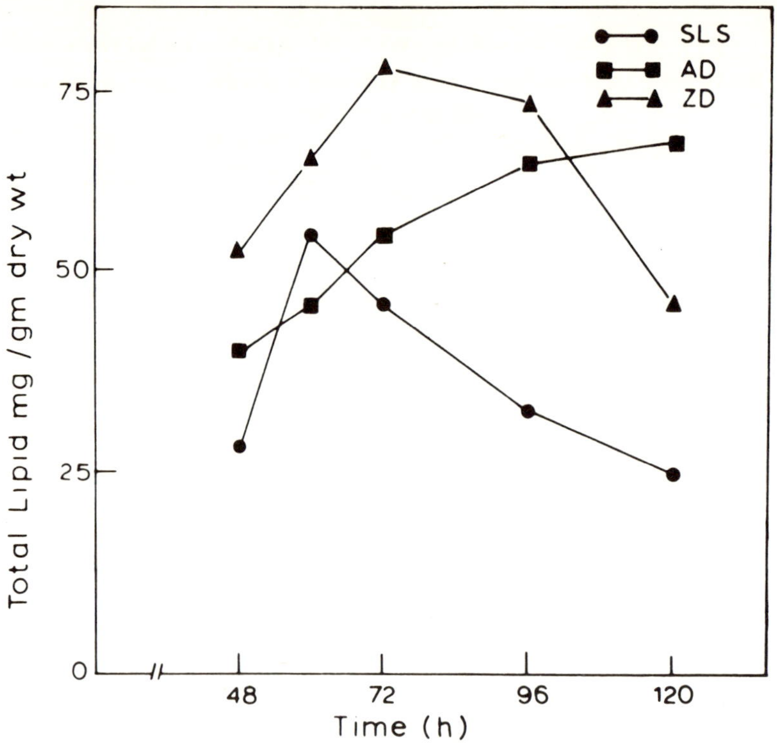

FIGURE 7. Total lipids in *A. parasiticus*

and lipids and also the intracellular concentration of pyridine nucleotides under different growth conditions. Thus, to gain a realistic assessment of the role of pyridine nucleotides in regulating the formation of lipids and aflatoxins, the experiments were conducted under the following different conditions:

1. Aflatoxins are produced in higher amount than lipids, e.g., SLS cultures.
2. Lipid content is more than aflatoxins, e.g., zinc-deficient (ZD) cultures.
3. Lipids and aflatoxins synthesized represent an intermediate level between SLS and ZD cultures, e.g., asparagine-deficient (AD) cultures.

The pattern of aflatoxin formation in the three different cultures has already been shown in Figure 1. Total lipids were also estimated and are depicted in Figure 7. It is evident from the figure that lipid levels follow a different pattern under different growth conditions. Under the conditions of reduced aflatoxin formation, e.g., AD and ZD cultures, maximum amount of lipid is present during the transitional growth phase (60 hr) and then decreases till stationary growth phase. One interesting fact that emerges from the study of levels of aflatoxins and lipids is that a correlation exists between the decrease in aflatoxin levels and a proportional increase in the lipid levels. In ZD cultures, the aflatoxin produced is minimum. In AD cultures, the aflatoxin amount is more than in ZD cultures, but less than in SLS cultures. Consequently, lipid synthesized in AD cultures is less than ZD cultures, but more than in SLS cultures.

The levels of NAD, NADP, NADH, and NADPH were assayed in SLS, AD, and ZD cultures, and are shown in Tables 4 to 6.[5] In all three cultures, NAD is the predominant oxidized nucleotide present at any stage of growth. In SLS and ZD cultures, the levels of NAD are high initially, and then decrease with age. The decline in NAD content is gradual

Table 4
LEVELS OF NAD, NADP, NADH, AND NADPH AT DIFFERENT GROWTH PERIODS IN *A. PARASITICUS* NRRL 3240 GROWN IN SLS MEDIUM

Period of growth (hr)	µmol/g dry wt					
	NAD	NADP	NADH	NADPH	NADH/NAD	NADPH/NADP
24	1.2	0.42	1.3	0.34	1.08	0.829
48	1.17	0.38	1.2	0.32	1.05	0.842
60	0.83	0.229	0.90	0.48	1.08	2.09
72	0.80	0.33	0.865	0.22	1.08	1.00
96	0.82	0.48	0.815	0.32	0.990	0.66
120	0.77	0.34	0.770	0.31	1.0	0.91

Note: *A. parasiticus* was grown as shake cultures at 28 ± 1°C for different growth periods. The mycelia were washed continuously. NAD, NADP, NADH, and NADPH were extracted and assayed. The values are the average of two separate experiments run in duplicate.

Table 5
LEVELS OF NAD, NADP, NADH, AND NADPH AT DIFFERENT GROWTH PERIODS IN *A. PARASITICUS* NRRL 3240 GROWN IN ASPARAGINE-DEFICIENT SLS MEDIUM

Period of growth (hr)	µmol/g dry wt					
	NAD	NADP	NADH	NADPH	NADH/NAD	NADPH/NADP
24	0.52	0.140	0.55	0.19	1.05	1.357
48	0.50	0.10	0.504	0.20	1.008	2.0
60	0.47	0.120	0.389	0.175	0.82	1.45
72	0.40	0.07	0.400	0.160	1.0	2.28
96	0.40	0.10	0.363	0.140	0.75	2.8
120	0.48	0.10	0.195	0.180	0.975	1.8

Note: Culture conditions were the same as described in Table 4. The values in the table are the average of two separate experiments run in duplicate.

Table 6
LEVELS OF NAD, NADP, NADH, AND NADPH AT DIFFERENT GROWTH PERIODS IN *A. PARASITICUS* NRRL 3240 GROWN IN ZINC-DEFICIENT SLS MEDIUM

Period of growth (hr)	µmol/g dry wt					
	NAD	NADP	NADH	NADPH	NADH/NAD	NADPH/NADP
24	0.921	0.246	1.62	1.11	1.75	4.51
48	0.844	0.230	1.5	1.0	1.77	4.34
60	0.350	0.180	0.5	0.7	1.42	4.0
72	0.420	0.130	0.65	0.58	1.54	4.46
96	0.360	0.100	0.56	0.35	1.55	3.5
120	0.630	0.100	0.81	0.43	1.28	4.3

Note: Culture conditions were essentially the same as described in Table 4. NAD, NADP, NADH, and NADPH were extracted and estimated. The values in the table are an average of two separate experiments run in duplicate.

in SLS cultures, while in ZD cultures the decline is comparatively sharp. In ZD cultures, a high level of NAD is observed at 120 hr. Under asparagine deficiency, the NAD levels fall gradually till 72 hr of growth and rise again at 96 hr. Under zinc and asparagine deficiency, levels of NADP fall continuously with age, while in SLS cultures a biphasic pattern is obtained. The NADP level in SLS cultures declines till the transitional growth phase and then increases again till stationary growth phase and finally drops down at 120 hr. The high levels of both the oxidized nucleotides correspond with the exponential growth phase. Fuhrer et al.[15] reported a decrease in NAD content with age for citric acid forming *Aspergillus niger* also. One main difference that has emerged from the study of levels of NAD and NADH in microorganisms is that the bacteria maintain a constant size of NADH pool, while eukaryotes maintain a constant NADH to NAD ratio.[16] We found that under normal conditions (i.e., SLS cultures) a constant ratio of NADH to NAD is maintained. Under the conditions of altered growth (e.g., AD or ZD cultures) this ratio is not maintained constant, suggesting a metabolic disorder.[5]

Among the reduced nucleotides, NADH is the predominant form. In all three cultures, the levels of NADH fall continuously with age. Under zinc and asparagine deficiency, the decrease in the level of NADH between exponential and transitional growth phase is rapid. In SLS cultures the NADH level falls gradually. High NADH levels correspond to high TCA cycle activity during the initial stages of growth.[17] A similar relationship between TCA cycle and NADH level has been demonstrated in *Mycobacterium phlei*.[18] It is interesting to note that the levels of reduced nucleotides are high under zinc deficiency. Elevation of levels of reduced nucleotides under zinc deficiency has been observed in other microorganisms also.[19] Another regulatory role served by NADH is the allosteric inhibition of malate dehydrogenase, citrate synthetase, malic enzyme, and phosphoenol pyruvate carboxykinase.[20] Among these enzymes, the activity of citrate synthetase has been monitored in our laboratory (Figure 6). Low activity of citrate synthetase was observed during intial stages of growth in SLS, AD, and ZD cultures. A spurt in the activity of citrate synthetase is observed at 72 hr in all three cultures. It is suggested that formation of citrate from oxaloacetate results in the dissipation of large amounts of free energy, thereby directing material from glycolysis into the citric acid cycle. The net result would be an accumulation of TCA cycle intermediates and finally the acceleration of aflatoxin biosynthesis. The inhibiton of phosphoenol pyruvate carboxykinase has been viewed as a mechanism to prevent the occurrence of gluconeogenesis when active glycolysis is on.[21] It is reasonable that malic enzyme and phosphoenol pyruvate carboxykinase are not essential when glycolysis is active because the products, pyruvate and phosphoenol pyruvate, are available from glycolysis.[20]

Higher NADH and NADPH levels may account for the higher lipid level in the ZD cultures. We have already dicussed the continued lipid synthesis under zinc deficiency (Figure 7). The observed low levels of NADH and NADPH are indicative of low overall metabolism. NADPH levels follow different patterns under different growth conditions. Under zinc and asparagine deficiency, its level declines gradually from early growth phase (24 hr) to late stationary growth phase. In SLS cultures, except for the variations during the active aflatoxin synthesis period (48 to 60 to 72 hr), the pool of NADPH is maintained at a fairly constant level. The highest level of NADPH is found during the transitional growth phase (60 hr), and then declines very sharply at early stationary growth phase (72 hr). It is possible that the reduced NADPH level under asparagine deficiency affects the system in such a way that the levels of all other nucleotides also decrease. That lowered NADPH level reduces the level of NADH has been demonstrated in *Neurospora crassa*.[21] In SLS cultures, a sharp burst in the NADPH level is observed during transitional growth phase, which reverts back to near normal value at 72 hr. It may be in response to the increased demand and comsumption of NADPH for aflatoxin synthesis. Of the two major NADPH-requiring pathways, the share for aflatoxins of NADPH seems to be higher during the later stages of growth, as is seen in the falling level of lipids (Figure 7).

Burton and Wilson[22] calculated the reduction potential of NAD and NADP and found that both have the same standard reduction potential. Hence, to understand the regulatory metabolic role played by pyridine nucleotides, regulatory charge values were calculated[5] and are shown in Tables 7 to 9. The charges measured are catabolic reduction charge, (CRC) defined as NADH/NAD + NADH, and anabolic reduction charge (ARC), defined as NADPH/NADP + NADPH. In addition, the ratio of oxidized vs. reduced nucleotides was calculated. It is clear from the data that the tendency of *A. parasiticus* is to maintain a constant CRC. The values of CRC reported for *A. parasiticus* are different from the values reported for the other molds (Table 10). In *A. parasiticus*, CRC, though constant for the particular set of conditions, varies in AD, ZD, and SLS cultures. In SLS and AD cultures, it is constant approximatley at 0.50, while under zinc deficiency, it is higher at 0.61.

A curious feature of catabolic reduction charge is that it remained constant over a wide range of intracellular concentrations of NAD and NADH. In other words, under conditions of excess or deficient pyridine nucleotides, cells appear to maintain this ratio rather than preserving the level of an individual component. CRC is a direct measure of the rate and direction of catabolic reactions in a cell. In SLS and AD cultures, the CRC is maintained at a comparatively lower value than in ZD cultures. The low CRC would pull NAD-linked reaction towards the delivery of reducing equivalents, as is characteristic of many catabolic reactions.[23]

The anabolic reduction charge in the three cultures follows a different pattern.[5] Under zinc deficiency, the ARC is maximum and is maintained constant during the different growth periods. In SLS cultures, a burst in ARC is observed during the transitional growth phase (60 hr), and then swiftly returns back to near normal value during the early stationary growth phase (72 hr). On the other hand, in asparagine deficiency the ARC value is higher than SLS but less than ZD. In AD cultures, also a burst in ARC value is observed, but in contrast to SLS the high ARC is observed during the exponential growth phase (48 hr). The burst in ARC value in AD cultures is of a small magnitude as compared to SLS cultures. A careful examination of the three media reveals that the ARC value is highest under zinc deficiency followed by AD and SLS cultures, respectively. A similar gradation with respect to lipid level has also been observed; the pattern is reversed with respect to aflatoxin content. Thus, it appears that a high ARC favors high lipid biosynthesis. Keiser and Niehaus[24] proposed that high ARC favors the flux of acetyl-CoA units into lipid synthesis over the polyketide formation. Our results provide experimental evidence to the postulate of Keiser and Niehaus.[24]

Two important parameters governing the ARC in the cell are (1) rate of protein synthesis and degradation[23] and (2) level of carbon and nitrogen source in the medium.[16] To find the reasons for the burst in the ARC value in AD (48 hr) and in SLS (60 hr) cultures, both of the above-mentioned possibilities were examined. Protein synthesis was followed by studying the uptake of ^{14}C-leucine by the mycelium at different stages of growth (Table 11).[5] Protein degradation was followed by assaying the protease activity during different growth periods in SLS, AD, and ZD cultures (Table 12). The uptake of leucine follows a biphasic pattern in SLS cultures. It reaches maximum during the exponential growth phase (48 hr), and then decreases till early stationary growth phase (72 hr). From 72 hr onwards, it rises again till 120 hr. The second peak in SLS cultures corresponds to the formation and germination of blastospores.[3] In AD cultures the maximum uptake occurs during the exponential growth period and then declines rather sharply from 72 hr till late stationary growth phase. In constrast, the maximum incorporation in ZD cultures takes place during the transitional growth phase and then decreases. As compared to AD cultures, the uptake of leucine in ZD cultures is more rapid during the entire stationary growth phase. The incorporation of leucine into the mycelium of different ages indicates that protein synthesis occurs at all stages of growth, though at a reduced rate during the stationary growth period. The control of ARC by the decline in protein synthesis can be ruled out since, in AD cultures, the ARC peak

Table 7
REDUCTION CHARGE AND OTHER DERIVED FUNCTIONS IN *A. PARASITICUS* NRRL 3240 AT DIFFERENT GROWTH PERIODS GROWN IN SLS MEDIUM

Period of growth (hr)	NADH + NAD	NADPH + NADP	(CRC) NADH/NAD + NADH	(ARC) NADPH/NADP + NADPH	(PNF) NADP + NADPH/NAD + NADH	NAD + NADP/ NAD + NADPH
24	2.5	0.75	0.52	0.453	0.30	0.98
48	2.408	0.70	0.514	0.450	0.29	0.99
60	1.730	0.709	0.520	0.67	0.40	0.72
72	1.665	0.550	0.510	0.40	0.33	1.04
96	1.638	0.80	0.490	0.40	0.48	1.206
120	1.540	0.650	0.500	0.47	0.42	1.027

Note: NAD, NADH, NADP, and NADPH were extracted and estimated. The values in the table represent an average of two separate experiments run in duplicate. CRC, catabolic reduction charge; ARC anabolic reduction charge; PNF, phosphorylated nucleotide fraction.

Table 8
REDUCTION CHARGE AND OTHER DERIVED FUNCTIONS IN *A. PARASITICUS* NRRL AT DIFFERENT GROWTH PERIODS GROWN IN ASPARAGINE-DEFICIENT SLS MEDIUM

Period of growth (hr)	NADH + NAD	NADPH + NADP	(CRC) NADH/ NAD + NADH	(ARC) NADPH/ NADP + NADPH	(PNF) NADP + NADPH/NAD + NADH	NAD + NADP/ NADH + NADPH
24	1.07	0.33	0.51	0.575	0.30	0.89
48	1.004	0.30	0.50	0.660	0.29	1.42
60	0.859	0.235	0.45	0.593	0.27	1.046
72	0.811	0.260	0.50	0.61	0.32	0.72
96	0.843	0.240	0.50	0.58	0.28	1.15
120	0.395	0.330	0.49	0.60	0.83	0.80

Note: NAD, NADH, NADP, and NADPH were extracted and estimated as mentioned in Table 4. The derived functions were then calculated. The values in the table represent an average of two separate experiments run in duplicate. CRC, catabolic reduction charge; ARC, anabolic reduction charge; PNF, phosphorylated nucleotide fraction.

Table 9

REDUCTION CHARGE AND OTHER DERIVED FUNCTIONS IN *A. PARASITICUS* NRRL 3240 AT DIFFERENT GROWTH PERIODS GROWN IN ZINC-DEFICIENT SLS MEDIUM

Period of growth (hr)	NADH + NAD	NADPH + NADP	(CRC) NADH/ NADH + NADH	(ARC) NADPH/ NADP + NADPH	(PNF) NADP + NADPH/NAD + NADH	NADH + NADP/ NADH + NADPH
24	2.541	1.350	0.637	0.818	0.53	0.42
48	2.344	1.23	0.63	0.81	0.52	0.42
60	0.850	0.86	0.58	0.80	1.05	1.106
72	0.070	0.58	0.60	0.816	0.66	0.44
96	0.920	0.40	0.60	0.777	0.48	0.50
120	1.44	0.830	0.58	0.811	0.36	0.58

Note: NAD, NADH, NADP, and NADPH were extracted and estimated as mentioned in Table 4. The derived functions were then calculated. The values in the table represent an average of two separate experiments run in duplicate. CRC, catabolic reduction charge; ARC, anabolic reduction charge, PNF, phosphorylated nucleotide fraction.

Table 10
PYRIDINE NUCLEOTIDES: LEVELS AND RATIOS IN DIFFERENT MICROORGANISMS

	Aspergillus niger[a]	*Escherichia coli*[b]	*Pseudomonas* sp.[c]
NAD	1.7 ± 0.25	8.6 ± 2.0	3.5
NADH	0.18 ± 0.03	1.1 ± 0.2	0.19 ± 0.05
NADP	0.14 ± 0.02	1.0 ± 0.2	0.27
NADPH	0.08 ± 0.02	1.3 ± 0.3	0.8 ± 0.18
CRC	0.10 ± 0.02	0.05	0.055 ± 0.0018
ARC	0.37 ± 0.05	0.57	0.74 ± 0.11
PNF	0.11 ± 0.02	0.19	0.22

[a] From Fuhrer et al.[15]
[b] From Andersen and Meyenberg.[23]
[c] From Matin and Gottschal.[16]

Table 11
UPTAKE OF ^{14}C-LEUCINE (DPM/MG DRY WT) BY THE MYCELIUM OF *A. PARASITICUS* NRRL 2340 AT DIFFERENT PERIODS OF GROWTH UNDER VARIOUS CONDITIONS

	Period of growth (hr)					
Medium used	**24**	**48**	**60**	**72**	**96**	**120**
SLS	621	1240	938	980	1028	1297
Asparagine deficient	748	1203	892	445	403	422
Zinc deficient	547	609	808	721	701	690

Note *A. parasiticus* was grown on SLS, AD, and ZD media for different periods at 28 ± 1°C as shake cultures. The total uptake of ^{14}C-leucine was then studied by resuspension experiments. The resuspension mixture consisted of 500-mg portion of mycelium, 5 mℓ of the respective media, and 1 μCi of ^{14}C-leucine. The suspension was incubated at 28 ± 1°C for 2 hr on a metabolic shaker. At the end of the incubation period, the uptake was stopped by adding 10 mℓ of 20% TCA. The mycelium was washed with 25 mℓ of distilled water to remove any adhering radioactivity. The washed mycelium was transferred to scintillation vial and dried at 50 ± 1°C. The radioactivity was measured in a Beckman Liquid Scintillation Counter (Model LS-3133). The values in the table represent an average of three separate sets of experiments run in duplicate.

occurs at 48 hr. In SLS cultures, though the decline in leucine uptake from 48 to 60 hr corresponds to ARC burst, the percentage decrease in leucine incorporation (approximately 25%) is too little to affect the ARC. In the case of *Escherichia coli* it has been demonstrated that complete inhibition of protein synthesis by chloramphenicol or rifampicin resulted in an increase in ARC.[24] Thus, the ARC burst due to protein synthesis is ruled out because (1) protein synthesis continues during the entire growth period, (2) the peak of ARC in AD cultures does not correspond with minimum uptake of leucine, and (3) the percentage decline in leucine uptake in SLS cultures (from 48 to 60 hr) is too little (25%) to account for the

Table 12
SPECIFIC ACTIVITY OF INTRACELLULAR PROTEASE IN *A. PARASITICUS* AT DIFFERENT GROWTH PERIODS UNDER VARIOUS CONDITIONS

	Period of growth (hr)					
Medium used	**24**	**48**	**60**	**72**	**96**	**120**
SLS	4.3	8.5	10.7	14.2	9.4	8.8
Asparagine deficient	12.9	15.1	21.65	22	32.8	38.9
Zinc deficient	6.0	9.9	15.1	22.9	16.1	17.0

Note: The mycelium was harvested and ground and protease was assayed in the cell-free extract. The values are expressed as tryptic units per milligram of protein and represent an average of two separate experiments run in duplicate.

ARC burst. Nevertheless, in SLS cultures the decline in leucine uptake may add synergistically with factor(s) in controlling the ARC level.

In addition to leucine uptake, protein degradation was also monitored.[5] Maximum protease activity was found during the early stationary growth phase (72 hr) in SLS and ZD cultures. In AD cultures, the protease activity increased continuously from exponential growth phase till late stationary growth phase. In SLS and ZD cultures, the proteolytic activity declined from early stationary growth phase. While in SLS cultures it accounts for the reversal of the mycelium intermediary growth pattern in ZD cultures, it may account for the continued lipid synthesis. The study of protease activity further substantiates that protein synthesis and degradation are not the major controlling factors for ARC regulation.

The study of depletion of carbon source (sucrose) and ammonium ions from the medium showed some interesting results (Figure 8).[5] Comparatively high levels of sucrose are left in the medium till the stationary growth phase and hence are not depicted. Under asparagine deficiency, since only ammonium sulfate as the nitrogen source is present, the rapid uptake of ammonium ions results in a sharp fall of ammonium ions concentration before the exponential growth phase. In SLS cultures, the lowest amount of ammonium ions is present during the exponential growth phase. In both cultures, the level of ammonium ions rises again. On the other hand, under zinc deficiency, the consumption of ammonium ions is very slow. It is indeed interesting to note that depletion of ammonium ions from the medium corresponds to the ARC in SLS and AD cultures. In SLS cultures, the minimum level of ammonium ions occurs at 48 hr, and ARC peak is at 60 hr; in AD cultures, the minimum value is at 30 hr, and the ARC peak is at 48 hr. Ammonium ion control of ARC is also evident from the fact that the magnitude of the peak in SLS or AD cultures corresponds with the rate of decline of ammonium ion levels in the two cultures. Ammonium ion control of ARC has been found in *Pseudomonas* species also. In addition, Matin and Gottschal[16] demonstrated that a decline in ammonium levels results in an increase in the intracellular pool of NADPH. We also observed a sharp increase in the NADPH pool at 60 hr of growth in SLS cultures. As for ARC, the decline in ammonia consumption can also account for the elevated NADPH pool in SLS medium. Thus, a sharp fall in extracellular ammonium ions in SLS and AD cultures results in an immediate burst in the ARC level. The ammonium ion regulation of ARC is further substantiated by the fact that, under zinc deficiency, the consumption of ammonium ions from the medium is gradual, and consequently no change

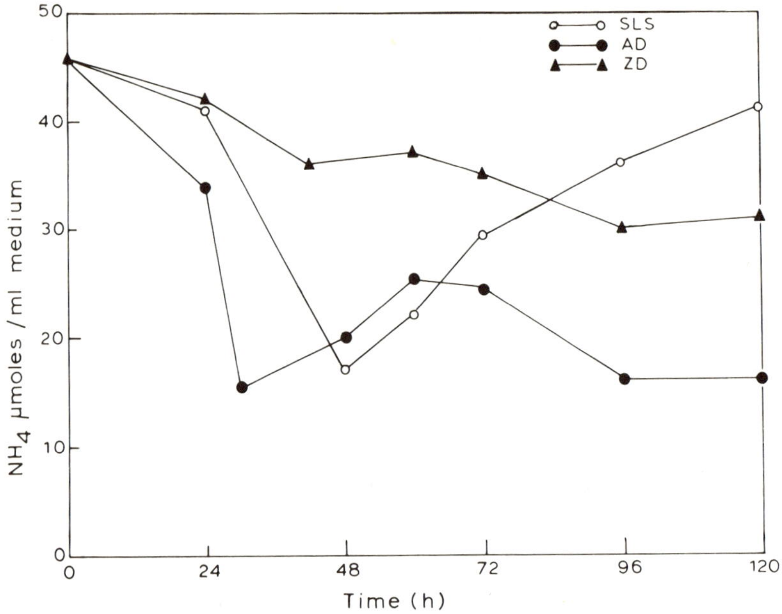

FIGURE 8. Level of ammonia in media of *A. parasiticus* at different hours.

in ARC is observed. The subsequent effect of ARC in SLS and AD cultures is different. In AD cultures, higher ARC favors the flux of acetyl-CoA preferentially into lipid synthesis; in SLS cultures and comparatively low ARC, the flux is more towards the synthesis of polyketide units. In ZD cultures a high level of ARC is maintained throughout the growth period and lipid synthesis also occurs at a higher rate.

IV. INVOLVEMENT OF POLYSUBSTRATE MONOOXYGENASE IN AFLATOXIN BIOSYNTHESIS

The biosynthesis of aflatoxins from acetate requires a series of oxidative transformation, which on general grounds is expected to involve polysubstrate monooxygenase (PSMO). The involvement of this enzyme system in the production of secondary metabolites has been demonstrated in several fungi, yeasts, and bacteria. Bu'Lock and Winstanley,[25] working on secondary metabolite formation in *Blakeslea trispora*, and Bu'Lock and Yuen[26] on *Trichoderma viride*, suggested a possible relationship between PSMO and secondary metabolite formation. The exact role of cytochrome P-450 in alkaloid biosynthesis in *Claviceps purpurea* was established by Ambike et al.[27] Subsequently, several reports on the possible role of cytochrome P-450 in a wide variety of reactions to secondary metabolism in fungi have been published. Breskvar and Hudnik-Plevnik[28] implicated cytochrome P-450 in 11α-hydroxylation of progesterone by *Rhizopus nigricans*. In the biosynthesis of patulin by *Penicillium patulum*, the hydroxylation of m-cresol to m-hydroxy benzylalcohol and 2,5-dihydroxy toluene was shown to NADPH and cytochrome P-450.[29] In fungi, simultaneous induction of secondary metabolism and cytochrome P-450 has been demonstrated in *Claviceps*[27] and implicated in *Trichoderma*.[26] We, in our laboratory, explored the possibile relationship between PSMO and aflatoxin biosynthesis.[30] To investigate this, we first carried out preliminary experiments in which we studied the effect of phenobarbitone, a known inducer of cytochrome P-450, on the aflatoxin production by *Aspergillus parasiticus*. The results obtained are shown in Table 13. It is evident that growth and aflatoxin production increased

Table 13
EFFECT OF PHENOBARBITONE ON AFLATOXIN PRODUCTION BY *A. PARASITICUS* NRRL 3240

Amount of phenobarbitone	Dry wt (g)	Aflatoxins (μg/100 mℓ) Mycelium $(B_1^B + B_2)$	Mycelium $(G_{12}^G + G_2)$	Medium $(B_1^B + B_2)$	Medium $(G_1^G + G_2)$	Total
Nil	3.0	1271	489	1501	435	3696
11	3.35	3642	1071	2783	601	8097
2	3.52	4938	1732	3168	706	10,544
5	3.20	2892	989	2071	531	6491
10	3.21	1489	672	1202	346	3709

Note: *A. parasiticus* was grown as shake cultures in 50 mℓ SLS medium containing different amounts of phenobarbitone for 96 hr. The mycelium was separated by vacuum filtration. Known amount of mycelium was ground with acid-washed sea sand and extracted three times with chloroform. Similarly, aflatoxins from medium were extracted three times with chloroform. The extracts were dried over anhydrous sodium sulfate, concentrated under vacuum, and made to a known volume. Aflatoxins were separated by silica gel G thin-layer chromatography using toluene:isoamyl:alcohol:methanol (90:32:3, v/v/v). The aflatoxin bands were visualized by UV light, scraped off the plates, and eluted three times with methanol. They were then estimated spectrophotometrically by recording their absorbances at 363 nm and using the respective extinction coefficients. The values represented are the average of two separate experiments run in triplicate.

with the increase in concentration of phenobarbitone. At high concentrations (5 and 10 mg/mℓ) both aflatoxin production and mycelial dry weight declined. It is clear that the increase in the mycelial aflatoxin content was more than the corresponding increase in the aflatoxin present in the medium. The maximum increase in total aflatoxin production was observed at 2 mg/mℓ of phenobarbitone. At this concentration, the total aflatoxin content increased by approximately 3.5-fold. It is also interesting to note that phenobarbitone decreased the ratio of aflatoxins to mycelium in medium. A similar increase in the production of another secondary metabolite—trisporic acid—by phenobarbitone was observed by Bu'Lock and Winstanley.[25] The observed increase in aflatoxin content by phenobarbitone suggested a possible involvement by PSMO in aflatoxin biosynthesis. Phenobarbitone is a known inducer of PSMO chain in prokaryotes and eukaryotes.[31] NADPH cytochrome P-450 reductase, an enzyme of PSMO chain, was assayed at different growth periods in SLS, AD, and ZD cultures.[30] The activity of NADPH cytochrome P-450 reductase and aflatoxin content at different hours is shown in Figure 9. The maximum activity of the enzyme is observed at different growth periods in the three cultures. In SLS cultures, maximum activity is observed during transitional growth phase (60 hr); in AD cultures, it is observed at 72 hr; in ZD cultures, it is observed at 48 hr. A common feature for all three cultures is that the maximum activity always precedes the intense aflatoxin synthesis period. Under asparagine and zinc deficiency, the maximum of NADPH cytochrome P-450 reductase is 35 and 60%, respectively, of those obtained in SLS cultures. The decrease in aflatoxin content in the three cultures also follows approximately the same pattern. Thus, a direct relationship between NADPH cytochrome P-450 reductase activity and aflatoxin biosynthesis is suggested. It stems mainly from the following two reasons:

1. Maximum aflatoxin production phase is preceded by a high NADPH cytochrome P-450 reductase activity.
2. The decrease in aflatoxin content under zinc and asparagine deficiency is proportional to the decrease in the NADPH cytochrome P-450 reductase activity.

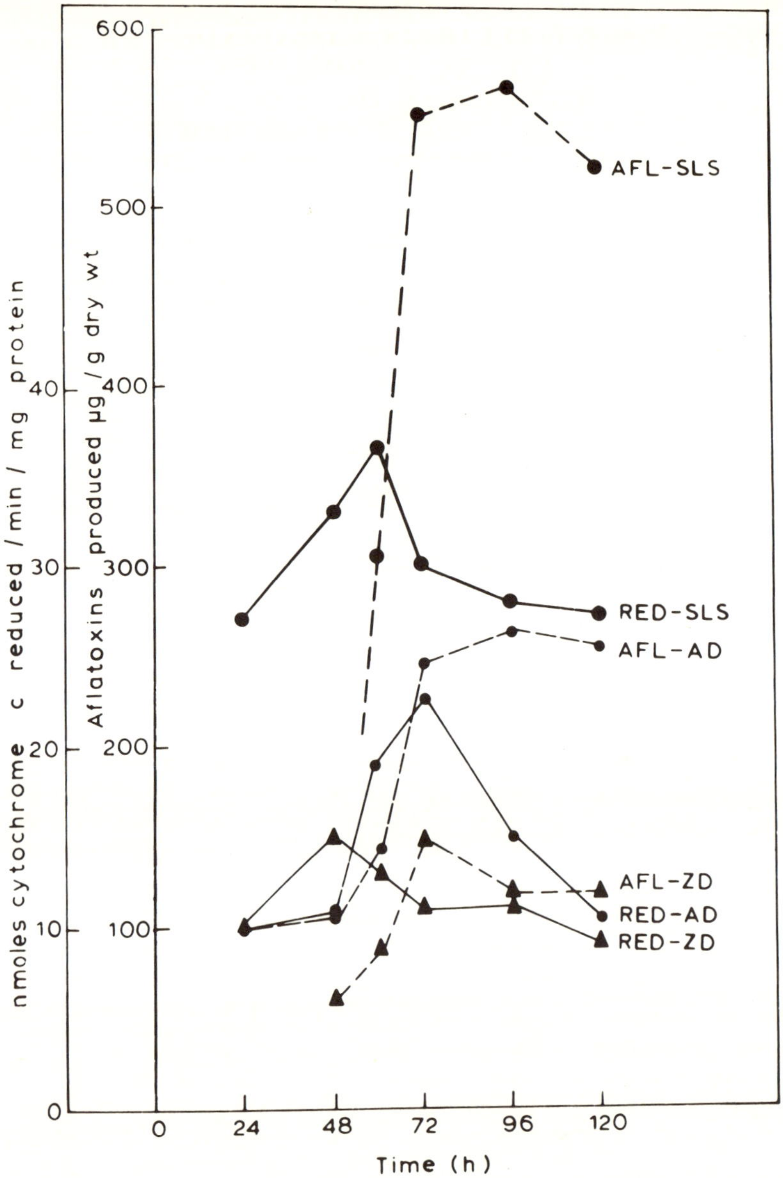

FIGURE 9. Aflatoxin production (μg/g dry wt) and NADPH cytochrome P-450 reductase. (From Bhatnagar, R. K., Ahmad, S., Kohli, K. K., Mukerji, K. G., and Venkitasubramanian, T. A., *Biochem. Biophys. Res. Commun.*, 104, 1287, 1982. With permission.)

Since NADPH cytochrome P-450 reductase is a membrane-bound enzyme, the microsomes of *A. parasiticus* were prepared and assayed for the enzyme. As is evident from Table 14, most of the enzyme activity was localized in the microsomes. It is interesting to note that the first report of cell-free synthesis of aflatoxins also implicated microsomes as the site of aflatoxin biosynthesis.[32] Phenobarbitone increased the microsomal NADPH cytochrome P-450 reductase by approximately threefold. This further substantiates the role of PSMO in aflatoxin biosynthesis. The microsomes were also able to metabolize benzphetamine, a known substrate for PSMO in the animal system (Table 15). The activity of PSMO towards benzphetamine was low. Benzphetamine is known to be a good substrate of rat liver cytochrome P-450.[33] Since various species of cytochrome P-450 are known to metabolize

Table 14
SUBCELLULAR LOCALIZATION OF NADPH CYTOCHROME P-450 REDUCTASE ACTIVITY IN *A. PARASITICUS* NRRL 3240 AND EFFECT OF PHENOBARBITONE ON ITS ACTIVITY

	Treatment	
Fraction	Control	Phenobarbitone (2 mg/mℓ)
12,000 × g supernatant	28.2	48.6
100,000 × g supernatant	10.6	21.7
Microsomes 100,000 × g sediment	44.2	116.0

Note: *A. parasiticus* was grown as described in Table 13. The reaction mixture in 1 mℓ contained phosphate buffer (pH 7.4), 500 μmol; cytochrome C, 0.1 μmol; EDTA, 5 μmol; NADPH, 1 mol and suitable amount of enzyme protein. Specific activity is expressed as nmol cytochrome *c* reduced/min/mg protein. The values reported are the average of two separate sets of experiments run in triplicate.

Table 15
SPECIFIC ACTIVITY OF BENZPHETAMINE N-DEMETHYLASE IN MICROSOMES OF *A. PARASITICUS* NRRL 3240

	Benzphetamine-*N*-demethylase	
Period of growth (hr)	Control (SLS)	PB (2 mg/mℓ)
48	0.750	—
96	2.520	3.28

Note: *A. parasiticus* was grown as described in Table 13. The reaction mixture in 1.5 mℓ contained Tris-HCl buffer (pH 7.5), 200 μmol; $MgCl_2$, 10 μmol; semicarbazide (pH 7.0), 5 μmol; benzphetamine, 5 μmol; NADPH-regenerating system and suitable amount of enzyme protein. After 15 min of incubation at 28 ± 1°C on a metabolic shaker, the reaction was stopped by $Ba(OH)_2$ and $ZnSo_4$. The tubes were centrifuged and formaldehyde formed was estimated in the supernatant. Specific activity is expressed as nmol formaldehyde formed/15 min/mg protein. The values reported are the average of two separate sets of experiments run in triplicate.

different substrates preferentially,[35] it is possible that benzphetamine is not an ideal substrate for fungal cytochrome P-450. Furthermore, the presence of phenobarbitone in the medium increased the PSMO activity towards benzphetamine by 32% (Table 15). These results were distinctly confirmed by the presence of PSMO in *A. parasiticus* and its involvement in aflatoxin biosynthesis.

To have a foolproof confirmation of the presence of cytochrome P-450 in *A. parasiticus*, microsomes were prepared on a large scale and solubilized. Cytochrome P-450 was isolated by ion-exchange chromatography on DE-52. Cytochrome P-450 was eluted by stepwise

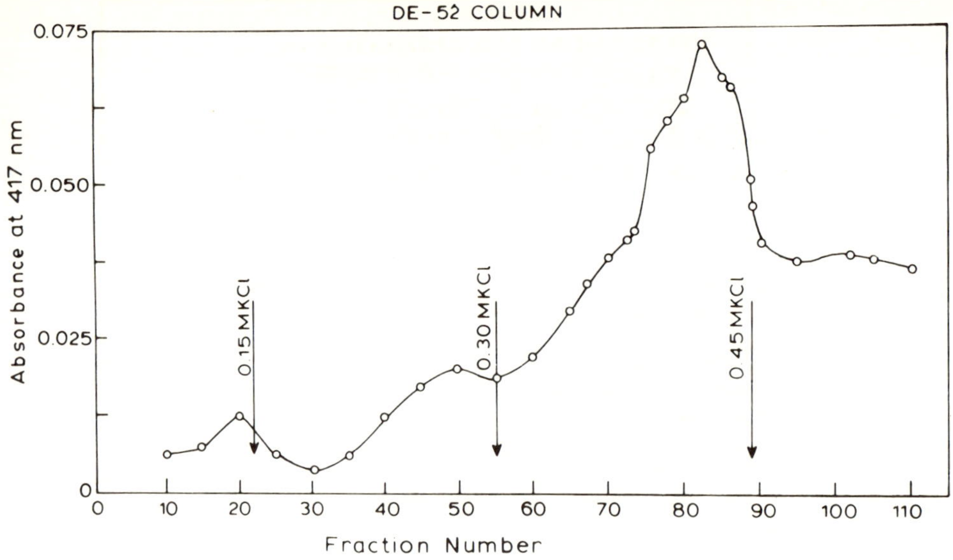

FIGURE 10. Ion-exchange chromatography of microsomes of *A. parasiticus* on DE-52. The microsomes were solubilized by Emulgen 911 and sodium cholate and the clear supernatant was chromatographed on DE-52 column (12 × 26 cm). The column was washed with 100 mℓ of 10 m*M* phosphate buffer (pH 7.4) containing 0.2% Emulgen 911, 0.5% sodium cholate, 0.1 m*M* EDTA, and 20% glycerol. Stepwise elution was carried out with the above buffer containing KCl. The column was operated at a flow rate of 30 mℓ/hr, and fractions were collected. The absorbance of the fractions was recorded at 417 nm.

gradient of potassium chloride.[5] The elution profile as monitored at 417 nm is shown in Figure 10. Oxidized and reduced spectrum of a fraction showing maximum absorbance at 417 nm was recorded. The shift in the spectral peak from 417 to 427.5 nm when reduced was suggestive of the presence of heme group in the active fractions (Figure 11).[5] The fractions showing absorbance at 417 nm were pooled and the assay of cytochrome P-450 performed. The carbon monoxide difference spectrum of reduced sample is shown in Figure 12. The peak with maxima at 446 nm establishes the presence of cytochrome P-450. The presence of peak at 424 nm in the carbon monoxide difference spectrum of reduced microsomes demonstrates the presence of cytochrome P-420. The presence of cytochrome P-420 itself supports the presence of cytochrome P-450, as cytochrome P-420 is a denatured form of cytochrome P-450. The specific activity of the isolated cytochrome P-450 was 0.25 nmol/mg protein. Since cytochrome P-420 represents the denatured form of cytochrome P-450, the specific activity of the isolated cytochrome P-450 can be taken as 0.57 nmol/mg protein.[30]

V. COENZYME-A POOL REGULATION OF AFLATOXIN BIOSYNTHESIS

CoA and its higher analogs occupy a central position in fungal metabolism. The formation of the malonyl-CoA units used for the biosynthesis of lipids and secondary metabolites, including aflatoxin, is generally presumed to be catalyzed by the same enzymes as forming a pool of biosynthetic intermediates. Biosynthesis of polyketides is known to require a series of condensation steps utilizing acetate and malonate units. Preference of malonate over acetate in the aflatoxin biosynthesis was demonstrated in our laboratory.[35] Mycelia of *A. parasiticus* were resuspended in different buffers and incorporation of 2-^{14}C malonate and 1-^{14}C acetate into aflatoxins was studied (Table 16). Mycelia took up comparatively more acetate than malonate, but more malonate (4.39%) was incorporated into aflatoxins than was acetate (1.5%). Furthermore, the addition of unlabeled acetate reduced the incorporation

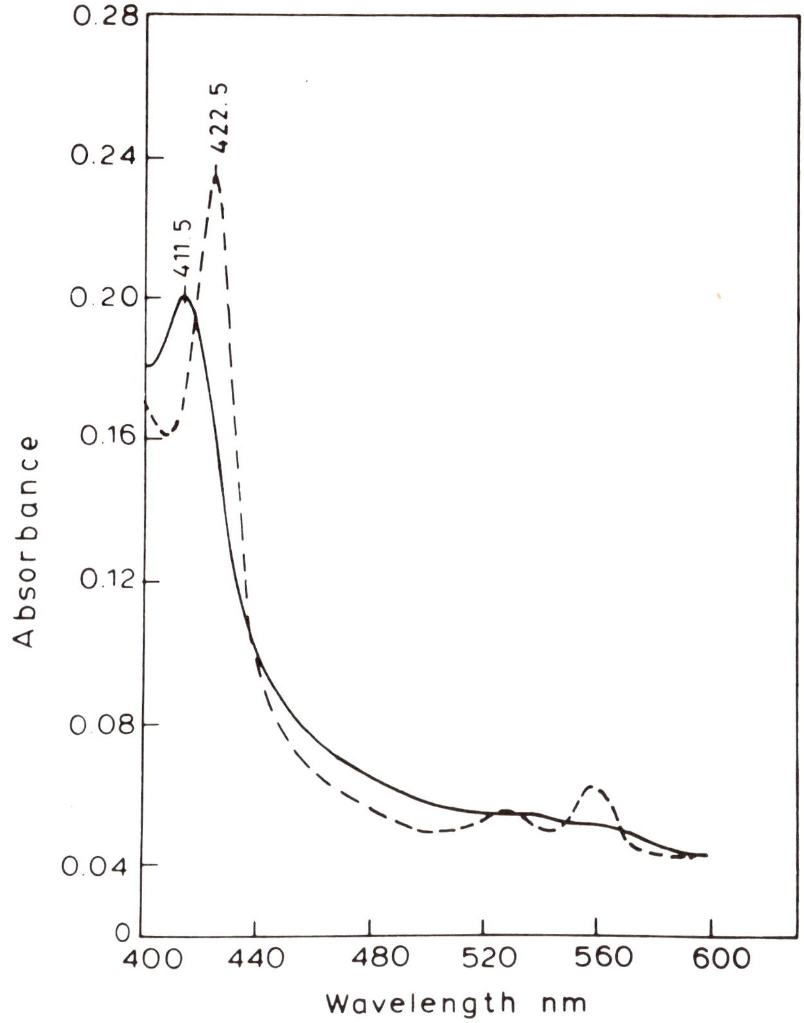

FIGURE 11. Absolute spectrum of cytochrome P-450 preparation from *A. parasiticus*. (————) Oxidized cytochrome P-450. (— — — — —) Reduced cytochrome P-450. P-450 was reduced by the addition of few milligrams of dithionite to the oxidized preparation. The maximum peak of oxidized preparation is observed at 415 nm and is shifted to 422.5 nm upon reduction. The spectrum was recorded on UV-VIS 210 Shimadzu.

of label from acetate (1-^{14}C) by 75%, but from 2-C^{14} malonate by only 25% (Table 17). This study was extended further and the coenzyme pools and enzyme systems involved in their metabolic pathways were investigated. Changes in the activities of the fatty acid synthetases, the two biotin-dependent carboxylases, acetyl-CoA and PEP carboxylases, and acetyl as well as acetoacetyl-CoA synthetases, were estimated before, during, and after the onset of aflatoxin biosynthesis.[11] The CoA pools of *A. parasiticus* under different growth conditions are given in Figures 13 to 15. The CoA levels (Figure 13) in SLS asparagine-deficient cultures reached a maximum at 72 hr of growth, while in zinc-deficient cultures they were maximum at 69 hr. The acetyl-CoA levels (Figure 14) in SLS cultures decreased with growth, while in asparagine- and zinc- deficient cultures they showed a continuous decrease, and in asparagine-deficient cultures it was maximum at 60 hr. However, a continuous increase was observed in SLS cultures. Acetyl-CoA synthetase activity shared a biphasic characteristic in all three media employed (Figure 16). Compared to activities in

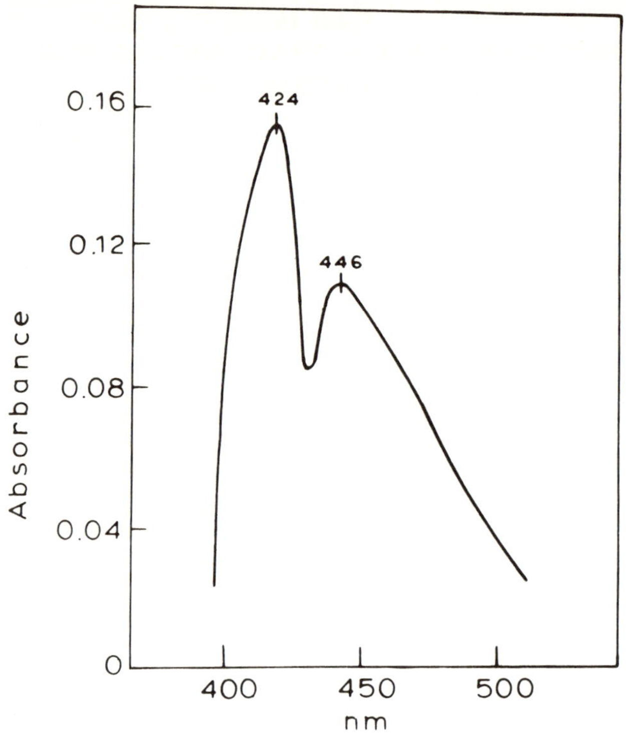

FIGURE 12. Carbon monoxide difference spectrum of the reduced cytochrome P-450 from microsomes of *A. parasiticus*. (From Lu, A. Y. H. and West, S. B., *Pharmacol. Rev.*, 31, 277, 1980. With permission.)

asparagine- and zinc-deficient cultures, the acetyl-CoA carboxylase activity was several times more in SLS cultures during the early growth phase (Figure 16). Under the deficient conditions it reached a maximum activity at 60 hr. The acetoacetyl-CoA synthetase activity (Figure 17) was high during the early growth phase, and declined with aging of the cultures in all three media used. The specific activity of fatty acid synthetase (Figure 18) was highest in SLS cultures during the early growth phase, and declined with growth subsequently. A similar pattern was observed in zinc-deficient cultures. However, under asparagine deficiency, maximum activity was observed at 60 hr. The specific activity of PEP carboxylase (Figure 18) showed different patterns under different culture conditions.

As already discussed, aflatoxin production was impaired under zinc and asparagine deficiency. The overall high lipid levels during the phase of intense aflatoxin production under zinc and asparagine deficiency indicate decreased availability of acetyl-CoA for acylation to form active precursors for aflatoxin synthesis. This probably accounts for the decreased production of aflatoxin under zinc and asparagine deficiency. The reduction in lipid levels in the SLS cultures is due to the intense production of aflatoxin, which competes for the same precursors. Since CoA is a direct precursor of aflatoxins and lipids, it is channelized into both of these pathways. In asparagine and zinc deficiency, where aflatoxin biosynthesis is impaired, elevated levels of acetyl-CoA were observed (Figure 14).

Malonyl-CoA, an intermediate in the biosynthesis of lipids and aflatoxins, increased in SLS cultures during the aflatoxin production phase (Figure 15). The high levels of malonyl-CoA observed in the early phase under deficient conditions where aflatoxin production is impaired rule out a probable role for malonyl-CoA in the induction of aflatoxin-synthesizing enzymes. It is pertinent to mention here that the results of the earlier experiments suggested that malonate was a direct precursor not only for aflatoxins, but also for lipids.

Table 16
INCORPORATION OF 1-^{14}C ACETATE AND 2-^{14}C MALONATE INTO AFLATOXINS B AND G

Resuspension medium	Compound added	Specific activity of aflatoxins (cpm/mg)	
		B	G
0.02 *M* phosphate buffer (pH 5.8)	(1-^{14}C) acetate	34,108	40,237
	(1-^{14}C) acetate + 40 m*M* EDTA	45,297	62,318
	(1-^{14}C) acetate + 40 mg unlabeled acetate	8,021	10,156
	(2-^{14}C) malonate	3,342	6,640
	(2-^{14}C) malonate + 40 m*M* EDTA	12,112	20,019
	(2-^{14}C) malonate + 40 mg unlabeled acetate	2,313	5,128

Note: *A. parasiticus* grown as stationary cultures were harvested on the 4th day and washed with double-distilled water. Known amounts were resuspended in 250-mℓ flasks containing 50 mℓ of sterile 0.02 *M* phosphate (buffer pH 5.8), citrate phosphate buffer (pH 2.8), or KCl-HCl buffer (pH 2.0), and 5 μCi (1-^{14}C) acetate or (2-^{14}C) malonate (specific activity 0.423 mCi/m*M*) was added. After incubating for 2 hr, aflatoxins were extracted with chloroform. Aflatoxins were eluted with methanol and estimated by spectrophotometry. Radioactivity incorporated into aflatoxins was counted on a Beckman LS-233 liquid scintillation counter.

Table 17
INCORPORATION OF (1-^{14}C) ACETATE AND 2-^{14}C MALONATE INTO MYCELIAL FRACTIONS

Fraction analyzed	Distribution of radioactivity (%)	
	(1-^{14}C) acetate	(2-^{14}C) malonate
Radioactivity remaining in resuspension buffer	38.8	88.4
$CHCl_3$ extract of mycelium	8.6	3.4
$CHCl_3$:MeoH (2:1, v/v) extract of mycelium	32.8	3.6
Residual mycelium	7.2	0.7
CO_2 evolved	3.0	0.5
Aflatoxins in the $CHCl_3$ extract	1.0	0.5

Note: A known amount of 4-day-old mycelium was divided into small portions and transferred to a Warburg flask containing 2.8 mℓ of 0.02 *M* citrate-phosphate buffer (pH 2.8) in the main compartment and 0.2 mℓ of 20% potassium hydroxide in the center well. (1-^{14}C) acetate or (2-^{14}C) malonate (0.5 μCi) was added from the side arm and the reaction allowed to proceed for 2 hr at 27°C. The trapped CO_2 was precipitated as barium carbonate and collected on filter paper discs. The mycelial portions were extracted first with chloroform and then with chloroform-methanol (2:1, v/v) and quantitatively transferred into scintillation vials. Radioactivity was measured in the original buffer, aflatoxins, ground mycelium, trapped CO_2, and mycelial extracts. Data presented are the averages of three experiments.

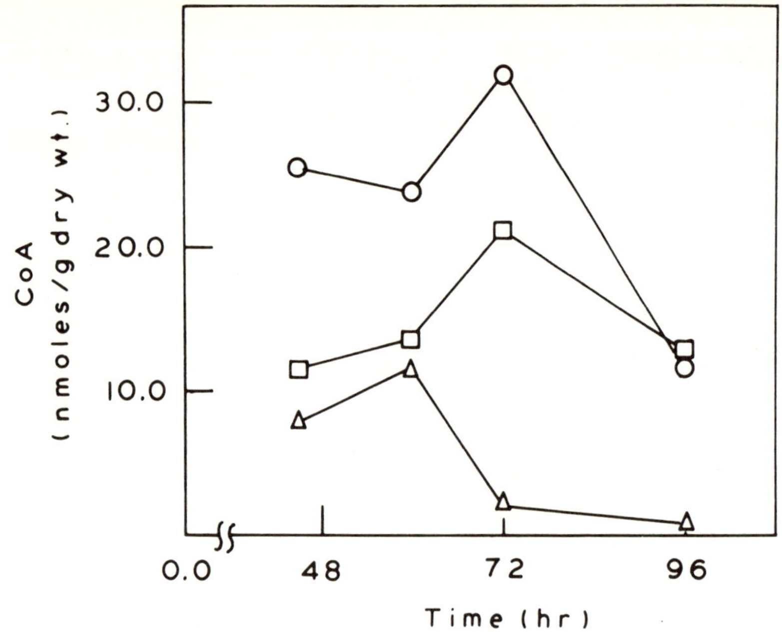

FIGURE 13. Intracellular CoA pool in *A. parasiticus* during aflatoxin production. (○ — ○) Sucrose-low salts (SLS) medium. (□ — □) Asparagine-deficient SLS (AD) medium. (△ — △) Zinc-deficient SLS (ZD) medium.

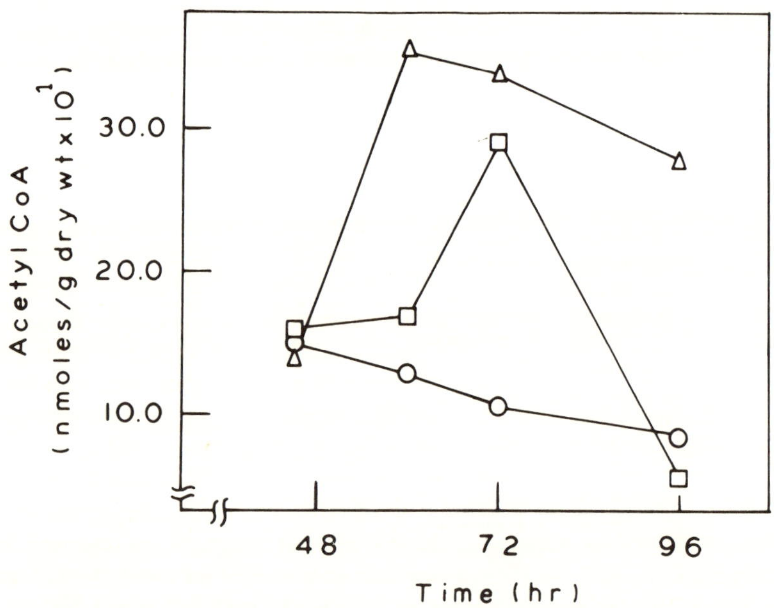

FIGURE 14. Intracellular CoA pool in *A. parasiticus* during aflatoxin production. (○ — ○) Sucrose-low salts (SLS) medium. (□ — □) Asparagine-deficient SLS (AD) medium. (△ — △) Zinc-deficient SLS (ZD) medium.

The specific activity of acetyl-CoA synthetase in *A. parasiticus* (Figure 16) is not in agreement with the levels of acetyl-CoA depicted in Figure 14. Similarly, no correlation was observed between acetyl-CoA synthetase and polifungin biosynthesis in *Streptomyces*

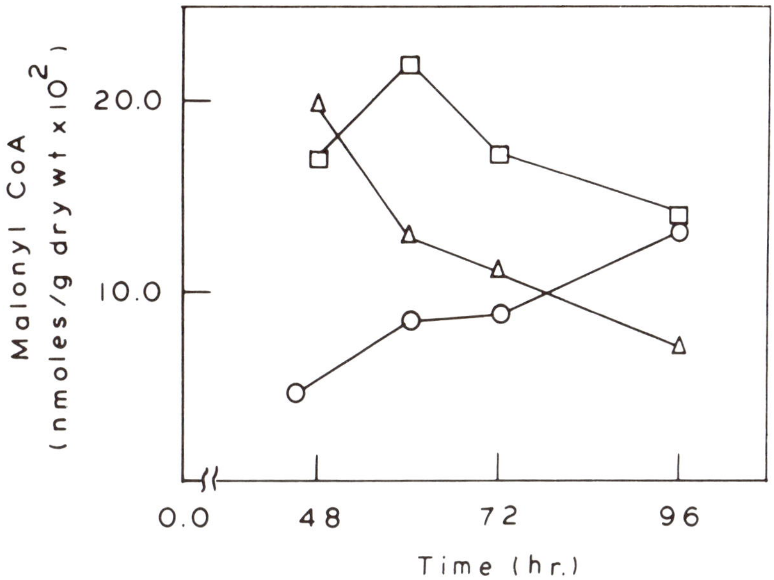

FIGURE 15. Intracellular CoA pool in *A. parasiticus* during aflatoxin production. (○ — ○) Sucrose-low salts (SLS) medium. (□ — □) Asparagine-deficient SLS (AD) medium. (△ — △) Zinc-deficient SLS (ZD) medium.

noursei var. *polifungini*.[36] The reason attributed to explain this anomalous observation states that acetyl-CoA is formed by several pathways.

Acetyl-CoA carboxylase, the key enzyme in fatty acid biosynthesis, had maximum activity in SLS cultures during the early growth period (Figure 16). Its specific activity declined during the early stationary growth phase. Alterations in the enzyme activity observed with aging in SLS cultures were not in agreement with the net levels of malonyl-CoA (Figure 15). The increase in net levels of malonyl-CoA during the period of decreasing acetyl-CoA carboxylase activity in the production phase probably indicates an alternate pathway for malonyl-CoA formation. Earlier studies have indeed shown the presence of an alternative pathway for malonyl-CoA formation in *Streptomyces aureofaciens*[37] and in *Penicillium islandicum*.[38] In both instances, an alternate pathway for the formation of malonyl-CoA via PEP carboxylase and an enzyme system decarboxylating oxaloacetate to malonyl-CoA besides the ''classical system'' by the carboxylation of acetyl-CoA were shown. We also observed[11] an increase in PEP carboxylase activity (Figure 18) during aflatoxin production phase in SLS cultures. This shows that the alternate pathway takes over from the classical system the production of malonyl-CoA, and thus accounts for the high malonyl-CoA levels during this period.

In zinc-deficient cultures, a similar pattern was observed in malonyl-CoA levels and enzyme activity. Though there is decreased aflatoxin formation, the accompanying high lipid synthesis at 96 hr may account for the low levels of malonyl-CoA observed. However, under asparagine deficiency, the PEP carboxylase activity declined with growth of the fungus, while acetyl-CoA carboxylase and the net malonyl-CoA levels followed a biphasic pattern.

The specific activity of acetoacetyl-CoA synthetase is high in the asparagine- and zinc-deficient cultures. This accounts for the increased lipid synthesis observed under these conditions as against those in SLS cultures. The highest activity of fatty acid synthetase coupled with that of acetyl-CoA carboxylase in SLS cultures during the early growth phase indicates a high primary metabolic activity in the fungus (Figure 17). It is interesting to note that as aflatoxin synthesis increased, the activity of fatty acid synthetase was reduced. The independence of fatty acid and secondary metabolite formation has been suggested for several microorganisms.[39]

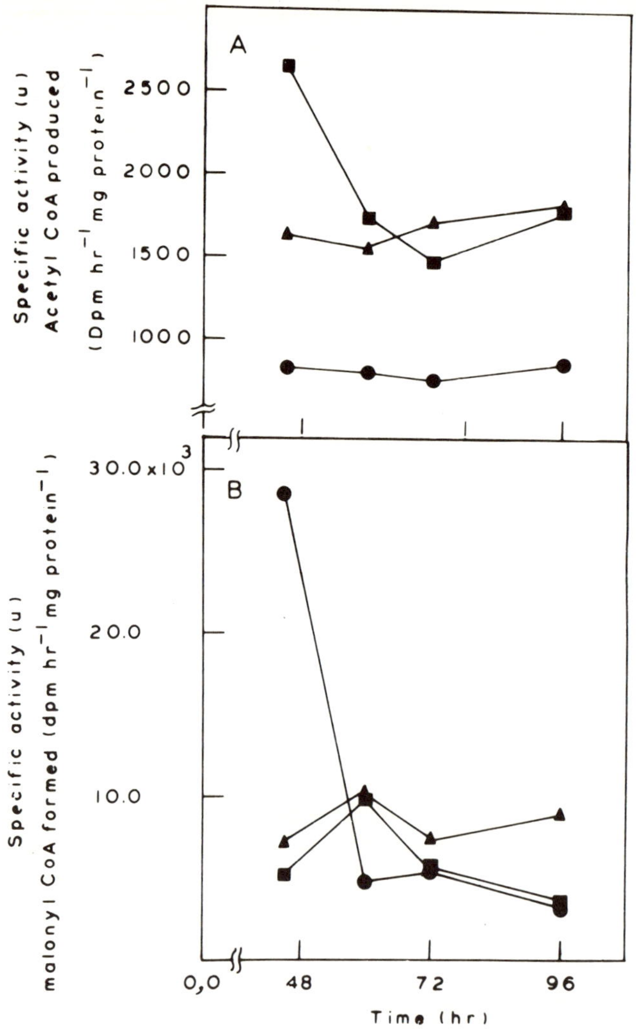

FIGURE 16. Changes in the activities of acetyl-CoA synthetase and acetyl-CoA carboxylase. (● — ●) Sucrose-low salts (SLS) medium. (■ — ■) Asparagine-deficient SLS (AD) medium. (▲ — ▲) Zinc-deficient SLS (ZD) medium.

VI. PREPARATION AND PROPERTIES OF SPHEROPLASTS FROM *ASPERGILLUS PARASITICUS* WITH SPECIAL REFERENCE TO THE *DE NOVO* SYNTHESIS OF AFLATOXINS

Attempts to prepare cell-free extracts active in *de novo* aflatoxin synthesis from *A. parasiticus* cultures by grinding with sea sand in the presence of sucrose/EDTA and mannitol/EDTA buffer proved unsuccessful. This suggested that the enzyme(s) involved in aflatoxin biosynthesis may be a part of a complex whose native conformation is altered by drastic treatments. It has been a matter of common observation that the conversion of mycelia to spheroplasts is followed by osmotic shock lysis to yield lysates which preserve the integrity of the enzyme complex. In our laboratory, we prepared spheroplasts of *A. parasiticus* using β-glucuronidase from *Helix pomatia*. The spheroplasts obtained were osmotically fragile spherical structures which lysed when suspended in hypnotic solutions (Figure 19).[40] Spher-

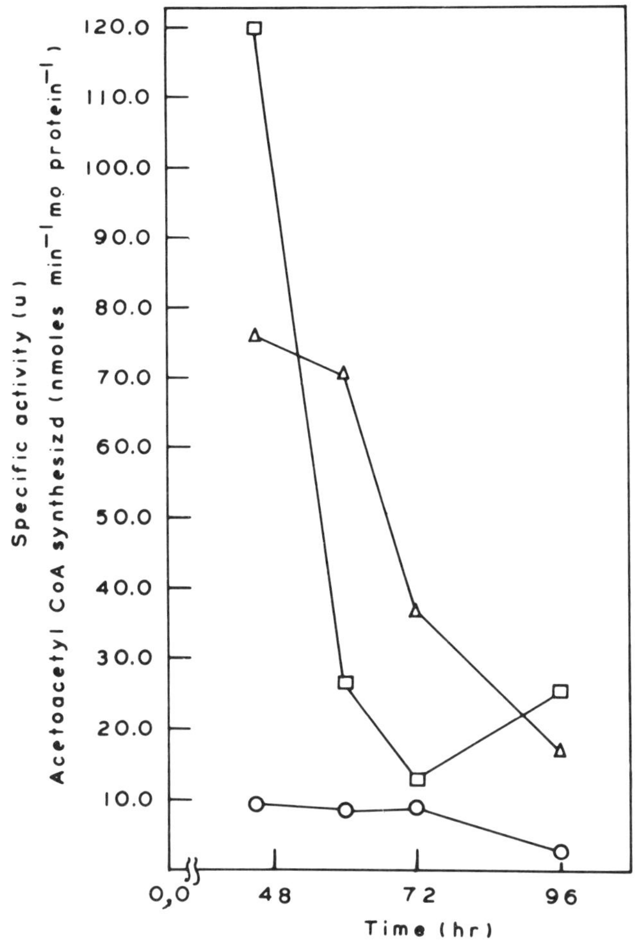

FIGURE 17. Changes in acetoacetyl-CoA synthetase activity during aflatoxin production. (○ — ○) Sucrose-low salts (SLS) medium. (□ — □) Asparagine-deficient SLS (AD) medium. (△ — △) Zinc-deficient SLS (ZD) medium.

oplasts were metabolically active as indicated by the incorporation of labeled thymidine, uridine, and leucine into DNA, RNA, and proteins, respectively (Figure 20).[41] Once the viability of spheroplasts was established, they were tested for their ability to incorporate labeled acetate into aflatoxins (Table 18). Spheroplast lysates also exhibited the ability to synthesize aflatoxins from labeled acetate (Table 18). Maximum incorporation of acetate into aflatoxins occurred spheroplasts and lysates from 96-hr cultures of *A. parasiticus*. In comparison to intact spheroplasts, lysates were observed to be more efficient in *de novo* aflatoxin synthesis from labeled acetate. Therefore, a comparison was made of the efficiency of 1-^{14}C acetate into aflatoxins by intact mycelia and spheroplast lysates prepared from them (Table 19). In Tris-HCL buffer, as compared to intact mycelia, lysates were nearly 45 times more efficient in aflatoxin synthesis from acetate on a protein basis as compared with mycelia. The lysates were almost 25 times more active in aflatoxin synthesis from labeled acetate in phosphate buffer. It is clear that lysates preserve the enzymes essential for the *de novo* synthesis of aflatoxins. Preparation of spheroplasts and lysates active in aflatoxin biosynthesis represents the initial step in the development of a cell-free system efficient in their *de novo* synthesis.

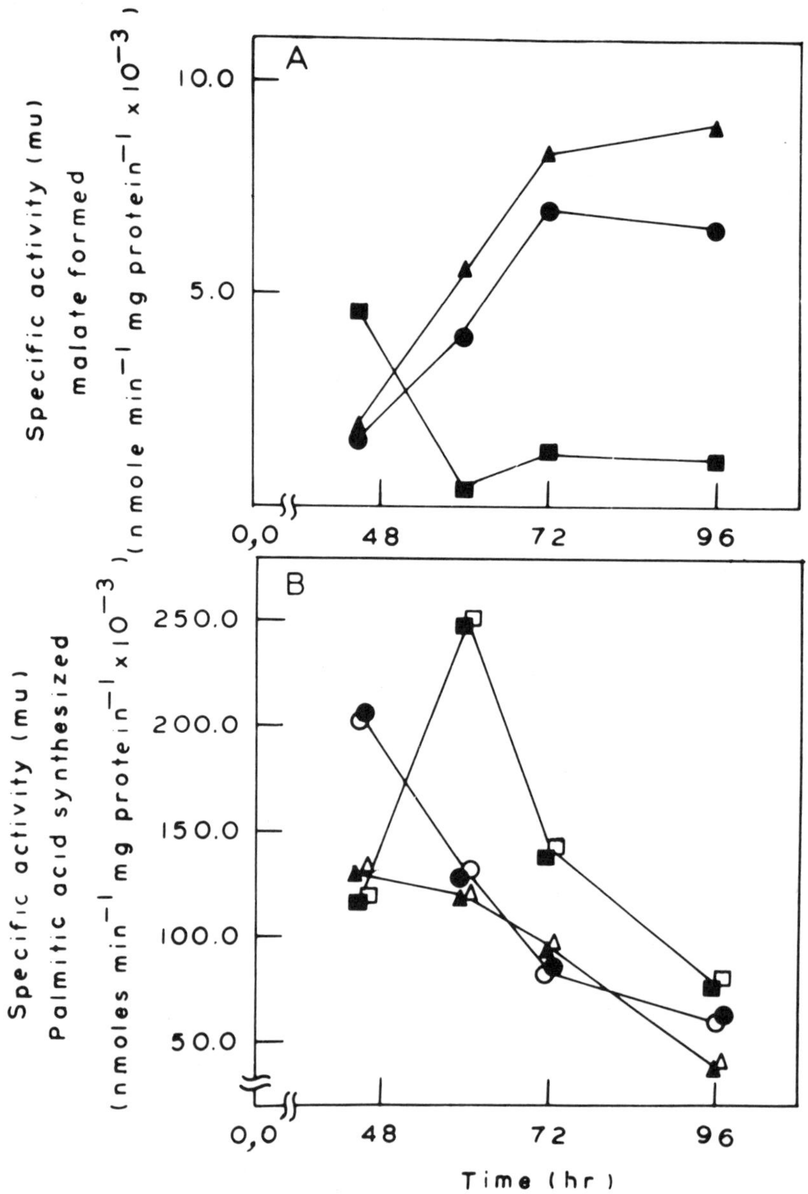

FIGURE 18. Variations in the activities of PEP carboxylase and fatty acid synthetase in *A. parasiticus*. (A) (● — ●) Sucrose-low salts (SLS) medium. (■ — ■) Asparagine-deficient SLS (AD) medium. (△ — △) Zinc-deficient SLS (ZD) medium. (B) (● — ●) Sucrose-low salts (SLS) medium. (□ — □) Asparagine-deficient SLS (AD) medium. (△ — △) Zinc-deficient SLS (ZD) medium.

VII. CONCLUSIONS AND FUTURE TRENDS

We have tried to project the metabolic regulation of aflatoxin biosynthesis with a basic view that a broad similarity exists between primary and secondary metabolism and secondary metabolism of one microorganism is quite similar to others. As in primary metabolism, secondary metabolism is also regulated by a narrow range of several regulatory parameters like energy charge, catabolic reduction charge, anabolic reduction charge, catabolite repression, and feedback inhibition.

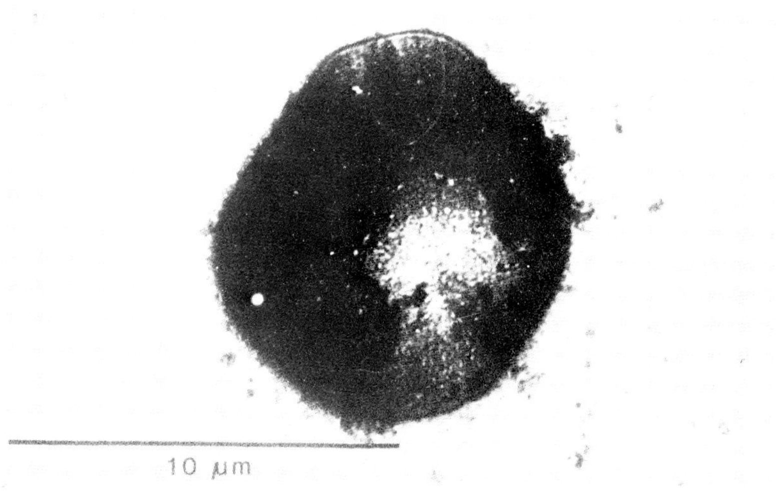

FIGURE 19. Electron micrograph of a single spheroplast after negative staining with phosphotungstic acid. (Magnification × 16,150.)

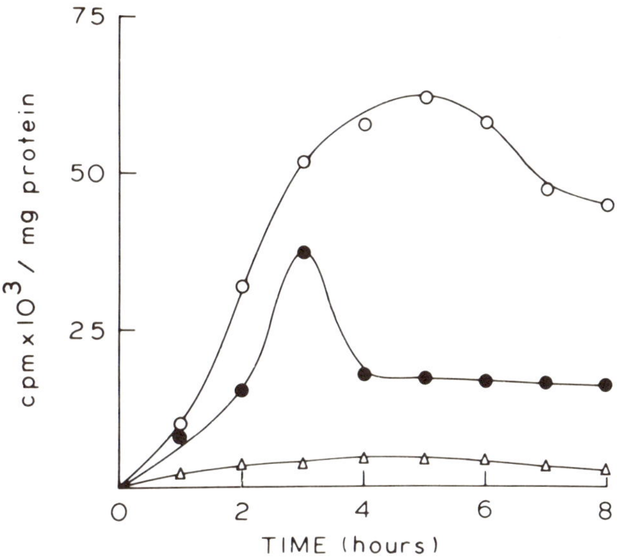

FIGURE 20. Macromolecular synthesis in spheroplasts. (○) Uridine. (●) Thymidine. (△) Leucine.

Table 18
INCORPORATION OF (1-^{14}C) ACETATE INTO AFLATOXINS BY INTACT SPHEROPLASTS AND SPHEROPLAST LYSATES

Age of culture (hr)	Incorporation of (1-^{14}C) acetate into aflatoxins (nCi/g protein ± S.D.)	
	Spheroplasts	**Lysates**
24	20.2 ± 2.3	222.0 ± 2.7
48	50.5 ± 3.6	356.7 ± 3.6
72	189.7 ± 3.6	389.0 ± 3.2
96	438.9 ± 8.2	1382.5 ± 18.2
120	76.7 ± 4.5	1144.3 ± 18.2

Note: Spheroplasts were prepared from *A. parasiticus* NRRL 3240 grown on SLS medium as shake cultures at 28 ± 1°C and suspended in 0.02 mol/ℓ phosphate buffer (pH 6.2) containing 0.05 mol/ℓ $MgSO_4$ and 0.6 mol/ℓ NaCl. For the preparation of the lysates, spheroplasts were suspended in 0.02 mol/ℓ phosphate buffer (pH 6.2). After incubation at 28 ± 1°C for 2 hr, radioactivity incorporated into aflatoxins was measured. The values reported are an average of six separate assays.

Although we now know about the control of aflatoxin biosynthesis, the enzyme system(s) responsible for their synthesis is not known. The inverse relationship between aflatoxin and lipid synthesis has been established beyond doubt, and it has been suggested that fatty acid synthetase during secondary growth phase becomes aflatoxin synthetase. Polysubstrate mon-

Table 19
COMPARISON OF THE INCORPORATION OF (1-^{14}C) ACETATE INTO AFLATOXINS BY MYCELIA, SPHEROPLASTS, AND SPHEROPLAST LYSATES

	Incorporation of (1-^{14}C) acetate into aflatoxins	
System	Suspending buffer	nCi/g protein ± S.D.
Mycelia	Tris-HCl (pH 7.2)	11.6 ± 2.0
	Phosphate (pH 6.2)	53.5 ± 6.0
Spheroplast lysates	Tris-HCl[a] (pH 7.2)	519.7 ± 5.0
	Phosphate[b] (pH 6.2)	1375.5 ± 13.1
Spheroplasts	Stabilizing buffer (pH 6.2)	430.3 ± 7.8

Note: Mycelia of *A. parasiticus* NRRL 3240, grown on SLS medium as shake cultures at 28 ± 1°C for 96 hr, were resuspended in 0.02 mol/ℓ Tris-HCl buffer (pH 7.2) and 0.02 mol/ℓ phosphate buffer (pH 6.2). Spheroplasts from the same cultures were similarly suspended in Tris-HCl and phosphate buffers. After incubation at 28 ± 1°C for 2 hr, radioactivity incorporated into aflatoxin was measured. The values are an average of six separate assays.

[a] These buffers cannot be used for testing the incorporation of (1-^{14}C) acetate into aflatoxins by intact spheroplasts, since they are hypotonic buffers.

[b] Phosphate buffer (0.02 mol/ℓ) containing 0.05 mol/ℓ $MgSO_4$ and 0.6 mol/ℓ NaCl.

oxygenases are strongly suspected to be involved in aflatoxin biosynthesis at certain stages at least. Whether the two complex enzyme systems operate in a concerted manner or whether only one system is involved is not exactly known. With the development of cell-free systems, spheroplast lysates capable of *de novo* aflatoxin biosynthesis, it should be possible to isolate active aflatoxin synthetase.

Some exciting work on the genetics of aflatoxin production has been done by Papa[42] and Bennet and co-workers,[43] but the identification of the gene(s) responsible for the synthesis of aflatoxins is still not known.

Aflatoxins are hydrophobic compounds, but fairly large quantities are secreted into the medium. We suspect that some binding factor is responsible for the secretion of aflatoxin from mycelium into the medium. Our preliminary experiments have indicated that the factor involved is a small molecular weight substance. However, the exact nature and the mechanism of its secretion are yet to be understood.

REFERENCES

1. **Gupta, S. K. and Venkitasubramanian, T. A.,** Production of aflatoxins on soya beans, *Appl. Microbiol.*, 29, 834, 1975.
2. **Detroy, R. W. and Hesseltine, C. W.,** Secondary biosynthesis of aflatoxin B_1 in *Aspergillus parasiticus*, *Can. J. Microbiol.*, 16, 959, 1970.
3. **Bhatnagar, R. K., Ahmad, S., Mukerji, K. G., and Venkitasubramanian, T. A.,** Role of blastospores in protecting *Aspergillus parasiticus* NRRL 3240 from high levels of aflatoxin, *Appl. Environ. Microbiol.*, 44, 579, 1982.
4. **Aharonowitz, Y. and Demain, A. L.,** Thoughts on secondary metabolism, *Biotechnol. Bioeng.*, 22(Suppl. 1), 5, 1980.
5. **Bhatnagar, R. K.,** Metabolism of *Aspergillus flavus* and *A. parasiticus* in Relation to Aflatoxin Production, Ph.D. thesis, University of Delhi, Delhi, 1982.
6. **Rao, V. B., Saraswathy, S., Maggon, K. K., and Venkitasubramanian, T. A.,** Bioenergetics of aflatoxin biosynthesis, *J. Food Sci.*, 45, 1031, 1980.
7. **Gupta, S. K., Maggon, K. K., and Venkitasubramanian, T. A.,** Effect of zinc on adenine nucleotide pools in relation to aflatoxin biosynthesis in *Aspergillus parasiticus*, *Appl. Environ. Microbiol.*, 32, 753, 1975.
8. **Atkinson, D. E.,** *Cellular Energy Metabolism and its Regulation*, Academic Press, New York, 1977.
9. **Knowles, C. J.,** Microbial metabolic regulation by adenine nucleotide pools, *Symp. Soc. Microbiol.*, 27, 241, 1977.
10. **Ward, A. C. and Packter, N. M.,** Relationship between fatty acid and phenol synthesis, *Eur. J. Biochem.*, 46, 323, 1974.
11. **Rao, V. B.,** Biosynthesis of Aflatoxins in *Aspergillus parasiticus*, Ph.D. thesis, University of Bombay, India, 1979.
12. **Shih, C. N. and Marth, E. H.,** Some cultural conditions that control the biosynthesis of lipid and aflatoxin by *Aspergillus parasiticus*, *Appl. Microbiol.*, 27, 452, 1974.
13. **Singh, R. and Hsieh, D. P. H.,** Enzymatic conversion of sterigmatocystin into aflatoxin B by cell-free extract of *Aspergillus parasiticus*, *Appl. Environ. Microbiol.*, 31, 743, 1976.
14. **Tyagi, J. S., Tyagi, A. K., and Venkitasubramanian, T. A.,** Some factors regulating (1-^{14}C) acetate incorporation into aflatoxins by spheroplasts and spheroplast lysates from *Aspergillus parasiticus*, *Toxicon*, 19, 445, 1981.
15. **Fuhrer, L., Kubicek, C. P., and Rohr, M.,** Pyridine nucleotide levels and ratios in *Aspergillus niger*, *Can. J. Microbiol.*, 26, 405, 1980.
16. **Matin, A. and Cottschal, J. C.,** Influence of dilution rate on NAD(P) and NADP(H) concentrations and ratios in a *Pseuodomonas* sp. grown in continuous culture, *J. Gen. Microbiol.*, 94, 333, 1976.
17. **Maggon, K. K., Gupta, S. K., and Venkitasubramanian, T. A.,** Biosynthesis of aflatoxins, *Bacteriol. Rev.*, 41, 822, 1977.
18. **Seshadri, R., Murthy, P. S., and Venkitasubramanian, T. A.,** Malate metabolism in *Mycobacterium*, *Biochem. J.*, 127, 62, 1972.
19. **Winder, F. G. and O'Hara, C.,** Effect of iron deficiency and of zinc deficiency on the composition of *Mycobacterium smegmatis*, *Biochem. J.*, 32, 198, 1962.
20. **Wright, J. A. and Sanwal, B. D.,** Regulatory mechanisms involving nicotinamide adenine nucleotides as allosteric efffectors. II. Control of phosphoenolpyruvate carboxykinase, *J. Biol. Chem.*, 247, 6013, 1969.
21. **Brody, S.,** Regulation of pyridine nucleotide levels and ratios in *Neurospora crassa*, *J. Biol. Chem.*, 247, 6013, 1972.
22. **Burton, K. and Wilson, T. H.,** The free energy changes for the reduction of diphosphopyridine nucleotide and the dehydrogenation of L-malate and L-glycerol-1 phosphate, *Biochem. J.*, 54, 86, 1953.
23. **Andersen, K. B. and Meyenberg, K.,** Charges of nicotinamide adenine nucleotides and adenylate energy charge as regulatory parameters of the metabolism in *Escherichia coli*, *J. Biol. Chem.*, 252, 4151, 1977.
24. **Keiser, R. C. and Niehaus, W. G.,** Purification and kinetic characterization of mannitol-l-phosphate dehydrogenase from *Aspergillus niger*, *Arch. Biochem. Biophys.*, 211, 613, 1981.
25. **Bu'Lock, J. D. and Winstanley, D. J.,** Trisporic production of *Blakeslea trispora* and its promotion by barbiturate, *J. Gen. Microbiol.*, 69, 391,1971.
26. **Bu'Lock, J. D. and Yuen, T. L. S.,** Oxygen requirements for secondary metabolism in *Trichoderma viride* and the effect of barbiturate, *Phytochemistry*, 10, 1835, 1971.
27. **Ambike, S. H., Baxter, R. M., and Zahid, N. D.,** The relation of cytochrome P-450 levels and alkaloid synthesis in *Claviceps purpurea*, *Phytochemistry*, 9, 1953, 1970.
28. **Breskvar, K. and Hudnik-Plevnik, T.,** A possible role of cytochrome P-450 in hydroxylation of progesterone by *Rhizopus nigricans*, *Biochem. Biophys. Biophys. Res. Commun.*, 74, 1192, 1977.

29. **Murphy, G., Vogel, G., Kripphal, G., and Lynen, F.,** Patulin biosynthesis: the role of mixed function oxidases in the hydroxylation of m-cresol, *Eur. J. Biochem.*, 49, 443, 1974.
30. **Bhatnagar, R. K., Ahmad, S., Kohli, K. K., Mukerji, K. G., and Venkitasubramanian, T. A.,** Induction of polysubstrate mono-oxygenase and aflatoxin production by phenobarbitone in *Aspergillus parasiticus* NRRL 3240, *Biochem. Biophys. Res. Commun.*, 104, 1287, 1982.
31. **Smith, R. V. and Davis, P. J.,** Induction of xenobiotic mono-oxygenases, *Adv. Biochem. Eng.*, 14, 62, 1980.
32. **Raj. H. G., Viswanathan, L., Murthy, H. S. R., and Venkitasubramanian, T. A.,** Biosynthesis of aflatoxins by cell-free preparations from *Aspergillus flavus, Experientia,* 25, 1141, 1969.
33. **Lu, A. Y. H. and West, S. B.,** Multiplicity of mammalian microsomal cytochrome P-450, *Pharmacol. Rev.*, 31, 277, 1980.
34. **West, S. B., Huang, M., Miwa, G. T., and Lu, A. Y. H.,** A simple and rapid procedure for the purification of phenobarbital inducible cytochrome P-450 from rat liver microsomes, *Arch. Biochem. Biophys.*, 193, 42, 1979.
35. **Gupta, S. R., Prasanna, H. R., Viswanathan, L., and Venkitasubramanian, T. A.,** Malonate as a precursor in the biosynthesis of aflatoxins, *J. Gen. Microbiol.*, 88, 317, 1975.
36. **Roszkowski, J., Kotiuszko, D., Rafalski, A., Morawska, H., and Raczynska Bojanowska, K.,** Characteristics of mutants of *Streptomyces noursei* var. *polifungini* producing antibiotics, *Acta Microbiol. Polom. Ser. B*, 4, 9, 1972.
37. **Behal, V., Jechova, V., Vanék, Z., and Hostalek, Z.,** Alternate pathways of malonyl-CoA formation in *Streptomyces aureofaciens, Phytochemistry,* 16, 347, 1977.
38. **Gatenbeck, S. and Mahlen, A. K.,** An alternate route for the formation of malonate in *Penicillium islandicum, Acta Chem. Scand.*, 22, 1696, 1968.
39. **Turner, W. B.,** Polyketides and related metabolites, in *The Filamentous Fungi. Biosynthesis and Metabolism,* Vol. 2, Smith, J. E. and Berry, D. R., Eds., Edward Arnold Press, London 1976, 445.
40. **Tyagi, J. S., Tyagi, A. K., and Venkitasubramanian, T. A.,** Preparation and properties of spheroplasts from *Aspergillus parasiticus* with special reference to the *de novo* synthesis of aflatoxins, *J. Appl. Bacteriol.*, 50, 481, 1981.
41. **Sivaswami, J.,** Role of Trace Elements in Production of Aflatoxins, Ph.D. thesis, University of Delhi, Delhi, 1978.
42. **Papa, K. E.,** Dominant aflatoxin mutant of *Aspergillus flavus, J. Gen. Microbiol.*, 118, 279, 1980.
43. **Bennet, J. W., Lee, L. S., Shoss, S. M., and Bondreaux, G. H.,** Identification of averantin as an aflatoxin B_1 precursor: placement in the biosynthetic pathway, *Appl. Environ. Microbiol.*, 39, 835, 1980.

Chapter 5

ROLE OF OXYGENASES IN THE METABOLISM OF PHENOLIC COMPOUNDS

C. S. Vaidyanathan

TABLE OF CONTENTS

I. INTRODUCTION

The various elemental cycles that operate in the biosphere are essential for the maintenance of life on this planet. Green plants trap the energy of the radiation from the sun to fix carbon dioxide into organic compounds. The carbon which is present in these compounds is transformed back into carbon dioxide by plants as well as animals, and the liberated energy is utilized for growth and maintenance of the living state. This dynamic sequence of events is represented by the carbon cycle.

The operation of a large number of anabolic pathways leads to the ultimate formation of a wide variety of substances of diverse structures. Several of these substances such as lignins, melanins, and tannins are phenolic in character and have to be biodegraded to maintain the carbon cycle. The biodegradation of these substances is achieved by soil microorganisms which have evolved versatile mechanisms to synthesize new enzymes. The catabolism of phenolic compounds is achieved both by respiration and by fermentation. In this review, further discussion will be restricted to the role of oxygenases in the metabolism of phenolic compounds.

II. BIOSYNTHESIS OF PHENOLS

A. Synthesis from Nonaromatic Precursors

All phenolic compounds are synthesized from acyclic precursors. These precursors are first converted to cyclic compounds like dehydroquinic acid, shikimic acid, and chorismic acid, which serve as intermediates in the biosynthesis of a large array of natural products. These include plastoquinones, plastochromanols, tocopherols, phylloquinones, menaquinones, ubiquinones, salicylic acid, pyrocatechuic acid, enterochelin, mycobactins, etc.,[1] of which several compounds are phenolic in nature. Some of the pathways leading to the biosynthesis of simple phenols are illustrated in Figure 1. The enzymology of several reactions of these pathways has not been studied in depth and offers ample opportunities for further studies.

An equally important mode of synthesis of phenols is through the acetate pathway. The polyketides obtained by the condensation of acetate units are produced mainly by fungi and to a lesser extent by higher plants. This group of compounds has no obvious function, but includes a number of antibiotics and plant pigments. Several of the salicylate derivatives, phloroglucinols, resorcylic acid derivatives, benzophenones, xanthones, flavonoids, isoflavonoids, rotenoids, stilbenes, naphthalenes, naphthoquinones, anthraquinones, tropolones, pyrromycinones, tetracyclines, and antibiotics such as griseofulvin, alternariol, penicillic acid, and patulin arise from acetate and related compounds.[2] Although the acetate pathway is equally important in the biosynthesis of a number of phenolic compounds, its enzymological aspects have not been as well understood as those of the shikimate pathway.

Details of the steps involved in the acetate pathway remained obscure for many years, but the recent work of Dimroth et al.[3,4] has elegantly illuminated the steps of the biosynthetic pathway. The biosynthesis of 6-methylsalicylate is one of the most intensively investigated areas of acetate pathway for the biosynthesis of aromatic compounds. 6-Methylsalicylate synthetase has been purified from *Penicillium patulum* and shown to be a multienzyme complex of molecular weight (M) 1.5×10^6. This enzyme resembles the yeast fatty acid synthetase system in several of its properties. By analogy with several other cyclization reactions, Lynen[5] has proposed a similar mechanism for the formation of 6-methylsalicyclic acid and extenced it to the synthesis of orsenillic acid (Figure 2).

Biosynthesis of gossypols, cannabinoids, and siccanochromes from terpenes and of estrogens and secosteroids[6-8] occurs through the operation of the isoprenoid pathway. In addition to these pathways, desaturation reactions also contribute to the formation of phenolic com-

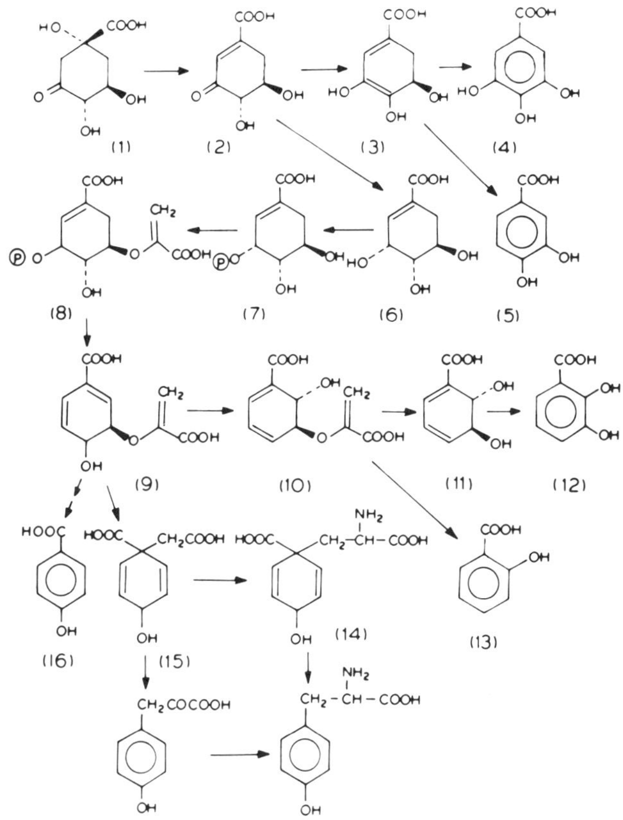

FIGURE 1. Biosynthesis of some simple phenolic compounds from dehydroquinic acid, shikimic acid, and chorismic acid.

pounds. An example of this is the conversion of 4-ketohexane-carboxylic acid to 4-hydroxybenzoic acid by the cell-free extracts of *Corynebacterium cyclohexanicum*[9].

1. Biosynthesis via the Action of Monooxygenases

Hydroxylation of the aromatic ring is catalyzed by monooxygenases. These enzymes are classified into six different groups: (1) pteridine-linked monooxygenases; (2) nicotinamide nucleotide-linked flavoproteins; (3) flavin-linked hemoproteins; (4) iron-sulfur-linked hemoproteins; (5) ascorbate-linked copper proteins; (6) monooxygenases that require another substrate as reductant. Since several excellent reviews[10,11] and references cited therein pertaining to these enzymes have appeared, only a brief discussion of the enzymology of these reactions is presented.

a. Pteridine-Dependent Monooxygenases

The conversion of phenylalanine to tyrosine is catalyzed by phenylalanine-4-hydroxylase. The purified enzyme from mammalian sources[12,13] requires tetrahydropteridine as the primary electron donor. Tetrahydropteridine is oxidized during the hydroxylation reaction to the quinonoid form of dihydropteridine, which is reduced back to tetrahydropteridine by dihydropteridine reductase. The sequence of reactions leading to the hydroxylation of phenylalanine is shown in Figure 3.

Extensive studies carried out by Kaufman[12] and Kaufman and Fisher[13] led to the characterization of enzyme-tetrahydropteridine-oxygen-substrate complex. This complex can undergo two reactions, one leading to hydroxylation and the other leading to the uncoupling

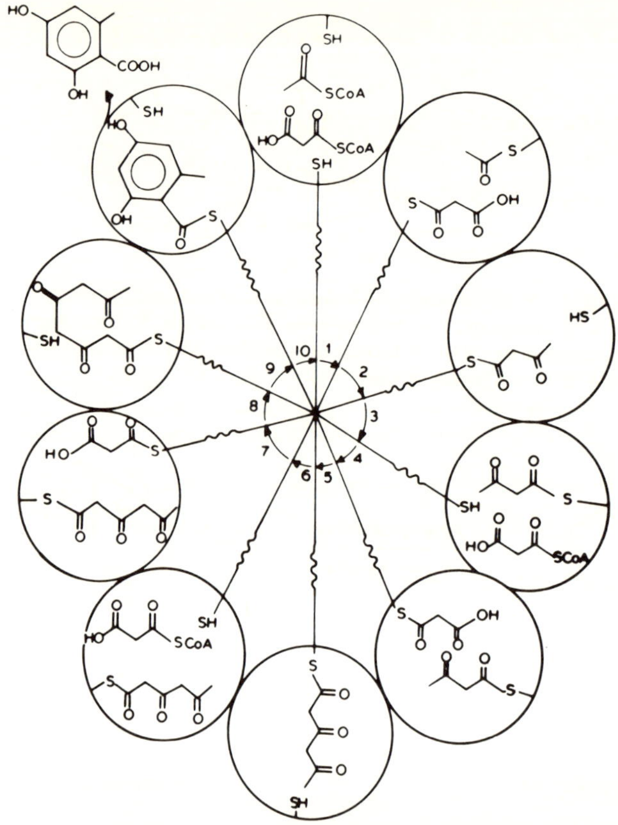

FIGURE 2. Hypothetical mechanism for the biosynthesis of orsenillic acid.

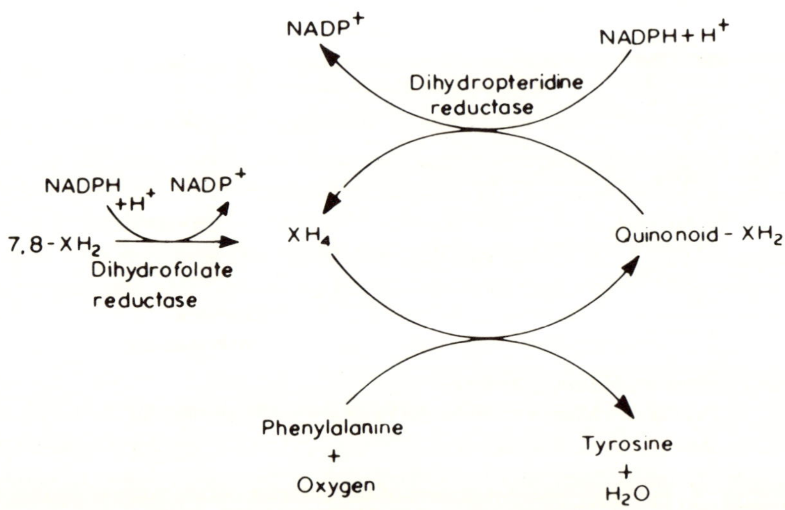

FIGURE 3. Scheme for the hydroxylation of phenylalanine. 7.8-XH_2 = 7.8-dihydropterine, XH_4 = tetrahydropterine.

$$E + O_2 + XH_4 + S \longrightarrow [E - XH_4 - O_2 - S]$$

$$[E - XH_4 - O_2 - S] \longrightarrow [E - XH_2 - O_2^{2-} - 2H^+ - S]$$

$$[E - XH_2 - O_2^{2-} - 2H^+ - S] \xrightarrow{(a)} E + XH_2 + S + H_2O_2$$

$$[E - XH_2 - O_2^{2-} - 2H^+ - S] \xrightarrow{(b)} E + XH_2 + SOH + H_2O$$

FIGURE 4. Reactions leading to hydroxylation or uncoupling of electron flow. a = uncoupling, b = hydroxylation, E = enzyme, S = substrate or pseudosubstrate, SOH = hydroxylated product, XH_4 = tetrahydropteridine, XH_2 = dihydropteridine

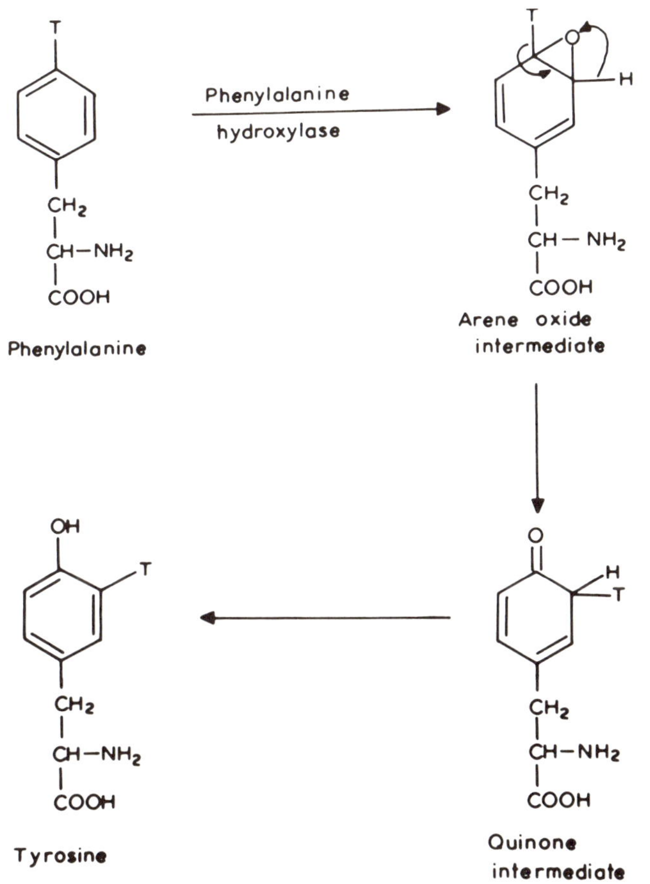

FIGURE 5. Mechanism for the NIH shift during phenylalanine hydroxylation.

of electron flow. In the latter case, oxidation of the cofactor occurs without hydroxylation, with the formation of hydrogen peroxide (Figure 4).

During the studies of phenylalanine hydroxylation, a phenomenon termed the ''NIH shift'' was discovered.[14] The migration of a *para* substituent to the adjacent *meta* position during hydroxylation of the 4-position of the benzene ring is called the ''NIH shift''. It has been proposed that during the NIH shift, an arene oxide and a quinone intermediate are formed, but this has not yet been confirmed (Figure 5).

In addition to the requirement for molecular oxygen and a pteridine cofactor, the role of iron in the catalysis has been elucidated by Fisher et al.[15] Unlike the mammalian phenylalanine hydroxylase, the enzyme from *Pseudomonas* species consists of a single polypeptide chain with a M_r 25,000.[16] The enzyme from *Chromobacterium violaceum* has also been found to be a monomer of 32,000 M_r, and no iron was detected in the pure enzyme.[17]

Recent studies showed that rat liver phenylalanine hydroxylase is a regulatory protein. The enzyme activity is modulated by substrate activation,[18-20] iron activation,[21] and phosphorylation and dephosphorylation.[22] The enzyme has also been shown to be activated by the modification of a single sulfhydryl group.[23] Studies on the interaction of pteridine cofactor and molecular oxygen with the enzyme revealed that neither carbon atom 6 or 7 of the pterin ring is involved in activation of oxygen.[24] Recent studies indicated the activation of oxygen via covalent addition of C(4a) tetrahydrobiopterin.[25]

Detailed studies on the role of pteridine in phenylalanine hydroxylase triggered the discovery of other pteridine-requiring enzymes such as tyrosine-3-hydroxylase and tryptophan-5-hydroxylase. Tyrosine-3-hydroxylase occurs in a variety of mammalian systems and is distinctly different from the tyrosinase which also catalyzes the conversion of tyrosine to 3,4-dihydroxyphenylalanine (DOPA).[13] Tyrosine hydroxylase resembles phenylalanine hydroxylase in several of its properties, such as pterin requirement, kinetic mechanism, etc.[13] Surprisingly, tyrosine hydroxylase catalyzes both phenylalanine hydroxylation reaction and tyrosine hydroxylation with equal efficiency. Recently, the presence of iron in the purified enzyme has also been established.[26]

Considerably less is known about the third pterin-dependent hydroxylase, namely tryptophan-5-hydroxylase. It specifically hydroxylates tryptophan at the 5-position and seems to be an iron-containing enzyme.[13] Phenylalanine is also hydroxylated by this enzyme. Letendre et al.[27] have characterized tryptophan hydroxylase from yet another source, i.e., *Chromobacterium violaceum,* which requires molecular oxygen, a reduced pteridine, and a sulfhydryl compound for its activity.

Benzoate-4-hydroxylase from a soil *Pseudomonas*[28] and from *Aspergillus niger*[29] has been isolated from our laboratory, and they were shown to be tetrahydropteridine-dependent enzyme. Both enzymes prefer NADPH as the other electron donor. While the fungal enzyme has a pH optimum of 6.2, the bacterial enzyme has a pH optimum of 7.2. Both enzymes required ferrous iron for activity. A number of substituted benzoic acids uncoupled the electron flow from hydroxylation in the bacterial system, in a manner similar to that observed in the case of mammalian phenylalanine hydroxylase.

A tetrahydropteridine- and iron-dependent enzyme which hydroxylates mandelic acid at the 4-position was reported from this laboratory.[30]

A phenylacetate-3-hydroxylase, which requires tetrahydropteridine for its activity, has been reported to occur in *Rhizoctonia solani.*[31]

b. Flavoprotein Monooxygenases

The flavoprotein monooxygenases form a fascinating group of enzymes, and the availability of highly purified crystalline enzymes has enabled many interesting studies to be undertaken. Flashner and Massey,[32] Massey and Hammerich,[33] and Massey[34] have reviewed the work on flavoprotein oxygenases in great detail. Hence, only a summary of these studies in relation to the recent findings is given below.

Salicylate hydroxylase is the first external flavoprotein monooxygenase to be characterized. Extensive studies carried out by Katagiri and group[35] on the enzyme from *Pseudomonas putida* led to the postulation of the mechanism shown in Figure 6. According to this scheme, the enzyme binds salicylate to form the ES complex, followed by the binding of NADH to give E-S-NADH complex. NAD^+ dissociates from the complex to give reduced enzyme-salicylate binary complex. The reduced enzyme-salicylate complex is also formed by the

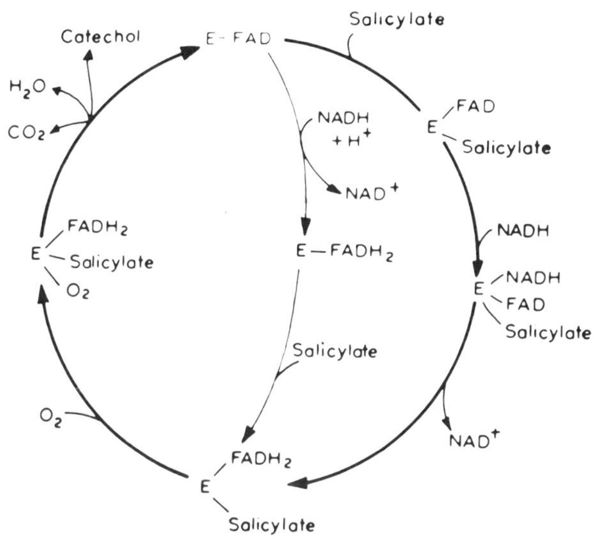

FIGURE 6. Reactions of salicylate hydroxylase.

operation of a minor pathway involving reduction of the free enzyme, followed by the binding of substrate to the reduced enzyme. Reduced enzyme-salicylate complex then reacts with oxygen to form reduced enzyme-substrate-oxygen complex, which breaks down to give the products and the free enzyme. The reduced enzyme also reacts with oxygen to give hydrogen peroxide and the free enzyme.

White-Stevens and Kamin[36] have isolated a soil bacterium which differs significantly in some of the properties from the *Pseudomonas* enzyme.[35] These include M_r, subunit composition, flavin content, and some kinetic properties. Salicylate hydroxylase reported by White-Stevens and Kamin[36] binds both benzoate and salicylate at the same site; also, both of these substrates stimulate the oxygen activation by the reduced enzyme. However, benzoate, being an unsuitable substrate, decomposes the reduced enzyme-benzoate-oxygen complex to H_2O_2 and benzoate.

p-Hydroxybenzoate-3-hydroxylase, which converts 4-hydroxybenzoate to 3,4-dihydroxybenzoate (protocatechuate), has been crystallized from four different species of *Pseudomonas*. The enzymes from all these sources are very similar in the strict requirement for NADPH, K_m for 4-hydroxybenzoate, inhibition at high substrate concentration, and FAD content. However, the M_r of these enzymes range from 65,000 to 93,000. Extensive studies carried out in Massey's laboratory have led to the detection of a C(4a)peroxodihydroflavin-enzyme-substrate complex and two other intermediates which have not been fully characterized,[33,37,38] The mechanism of hydroxylation reaction proposed by Massey is shown in Figure 7. Detailed studies on the kinetic mechanism of this enzyme with 4-mercaptobenzoate as the substrate revealed that the first half of the reaction proceeds in a manner analogous to that with the natural substrate, while in the second half, 4-mercaptobenzoate is rapidly oxidized to give 4,4,′-dithiobisbenzoate, instead of being converted to the expected product, 3-hydroxy-4-mercaptobenzoate.[39]

Several recent papers by Massey's group highlight the usefulness of modified flavins such as 8-chloro, 8-mercaptoflavins,[40-42] 2-thio FAD,[43] and 1-deaza FAD [44] in the mechanistic studies on flavoproteins. It has been shown that 2-thio FAD in the presence of H_2O_2 forms a covalent linkage with some group at the active site of *p*-hydroxybenzoate hydroxylase.[45,46] Shoun et al. have studied the interaction of *p*-hydroxybenzoate and NADPH with the enzyme.[47,48] These authors have also identified an arginine at the aromatic substrate binding site and a histidine residue at the NADPH binding site of the enzyme.[49,50]

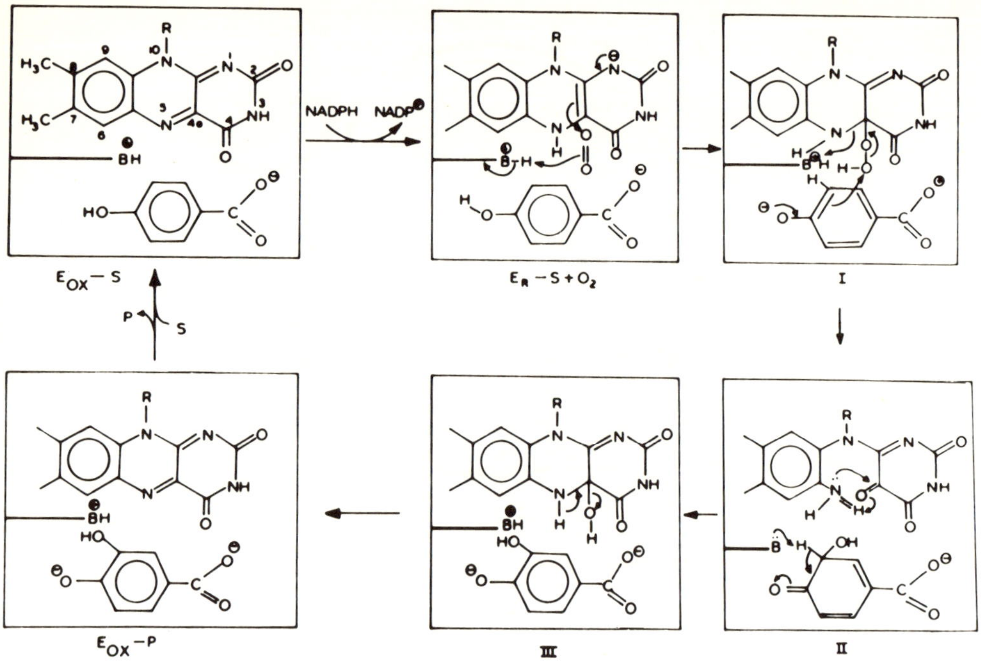

FIGURE 7. Proposed mechanisms of action of 4-hydroxybenzoate-3-hydroxylase.

Melilotate hydroxylase catalyzes the conversion of melilotate into 2,3-dihydroxyphenylpropionate with the incorporation of molecular oxygen and oxidation of NADH. The enzyme was first reported by Levy[51] in an *Arthrobacter* species and later extensively characterized by Strickland and Massey[52] from a *Pseudomonas* species. Although the enzyme from both sources contains FAD as the prosthetic group, the *Pseudomonas* enzyme is a tetrameric protein, while the *Arthrobacter* enzyme is a monomer. Fast reaction kinetic analyses revealed the operation of the mechanism shown in Figure 8. Strickland et al.[53] have also studied the possibility of NIH shift during melilotate hydroxylation using 3,5-dideuteromelilotate as substrate. They were unable to detect any NIH shift as there was no retention of deuterium. The absence of NIH shift does not rule out the arene oxide pathway, since it has been shown that substituents on phenol show negligible or no migration during aromatic hydroxylations with systems known to give the NIH shift with other substrates.[54]

Ribbons and associates[55,56] have characterized orcinol and resorcinol hydroxylases from *P. putida.* These functionally homologous enzymes are involved in the conversion of 1,3-dihydroxybenzene into 1,2,4-trihydroxybenzene. Both enzymes have been demonstrated by these investigators to be flavoprotein monooxygenases containing approximately 1 mol of FAD per mole of enzyme of M_r about 70,000. They are monomeric in nature and are two different enzymes as evidenced by the differences in circular dichroic spectra and utilization of various aromatic substances as substrates and effectors. This is an interesting example of functionally similar enzymes with considerable homology in the antigenic determinants, but with different three-dimensional structures.

Kynurenine hydroxylase which converts kynurenine to 3-hydroxykynurenine utilizing molecular oxygen and either NADH or NADPH, has been shown to be a FAD-containing monooxygenase based on reconstitution experiments. However, further studies are essential to identify the flavin cofactor.[32,33]

Neujahr and Gaal[57,58] have characterized a phenol hydroxylase from *Trichosporon cutaneum* which converts phenol to catechol utilizing NADH and oxygen. The enzyme has a M_r of 148,000 and contains approximately 2 mol of FAD per mole of enzyme. This enzyme

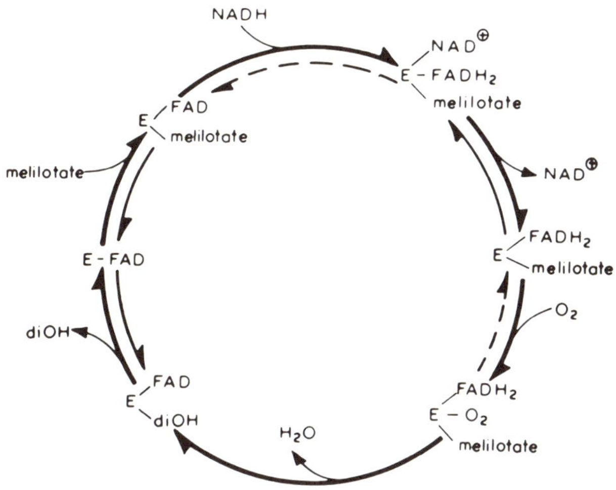

FIGURE 8. Mechanism of the reaction catalyzed by melilotate monooxygenase. diOH, dihydroxyphenylpropionate.

hydroxylates a number of phenols and is inhibited by heavy metal ions and 4-hydroxymercuribenzoate (pHMB). Like melilotate hydroxylase and 4-hydroxybenzoate hydroxylase, phenol hydroxylase also forms a transient peroxydihydroflavin intermediate when the reduced enzyme-phenol complex is mixed with oxygen.[33]

m-Hydroxybenzoate-4-hydroxylase from *P. testosteronii*[59] and 3-hydroxybenzoate-6-hydroxylase from *P. aeruginosa*[60] have been purified to homogeneity and characterized to be flavoprotein monooxygenases by Ribbons and Groseclose. From *Aspergillus niger,* 3-hydroxybenzoate-4-hydroxylase was purified in our laboratory and shown to be a NADPH-dependent flavoprotein monooxygenase.[61,62] Ashok Kumar et al.[63] suggested the participation of superoxide anion (O_2^-) in the reaction catalyzed by this enzyme.

Elmorse and Hopper[64] have purified a flavoprotein hydroxylase from *P. putida* grown on 2,4–xylenol. This enzyme, which can function either with NADH or NADPH, oxidatively decarboxylates 4-hydroxyisophthalate to give protocatechuate. This reaction is analogous to the salicylate hydroxylase reaction. The enzyme contains 1 mol of FAD per mole of enzyme of M_r 103,000.

Soon after the initial demonstration of the conversion of 4-hydroxyphenylacetate to homogentisate by Blakley,[65] Dagley's group[66] purified this enzyme and demonstrated the specific requirement for both FAD and Mg^{2+}. It is the first example of a flavoprotein monooxygenase showing the NIH shift, although the flavoprotein nature of the enzyme remains to be more rigorously characterized.

c. Cytochrome P-450 Monooxygenases

Cytochrome P-450 monooxygenases were recognized more than a decade ago as a special class of hemoproteins implicated in microsomal hydroxylations. Such proteins occur widely distributed in animals, plants, and microorganisms and play a very significant role in (1) the biosynthesis of several essential metabolites such as fatty acids, steroids, prostaglandins, etc., (2) the oxygenation of xenobiotic compounds and (3) the recycling processes in the biosphere carried out by several microorganisms.[67]

There are two types of cytochrome P-450 monooxygenases, one linked to the iron-sulfur proteins and the other linked to flavoproteins.

Camphor-5-exomonooxygenase from *P. putida,*[68] 11-β-steroid monooxygenase from adrenal mitochrondia, and cholesterol sidechain cleaving enzyme, belong to the first category,

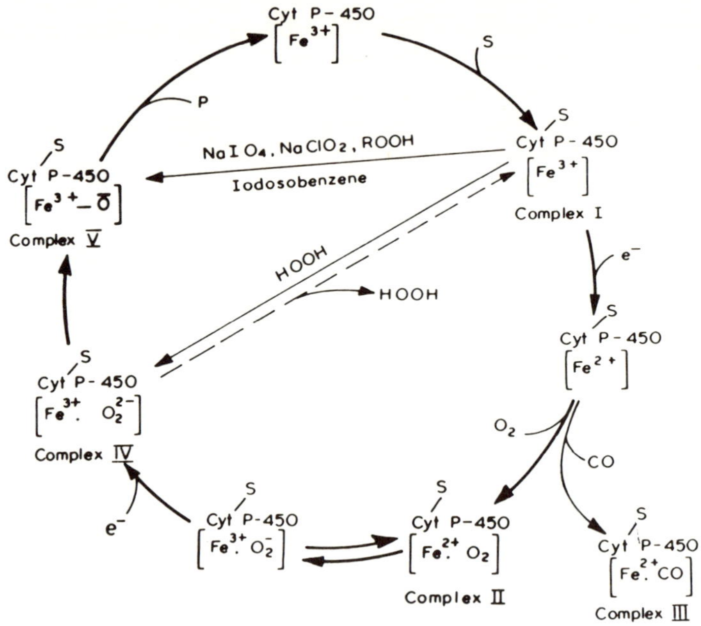

FIGURE 9. Mechanisms of action of cytochrome P-450 monooxygenases.

while the monooxygenase systems for xenobiotic compounds belong to the second category.[69] The latter enzyme system is involved in the conversion of a number of aromatic ompounds to phenolic derivatives. The aromatic hydroxylations catalyzed by this enzyme system are usually associated with the occurrence of the NIH shift (Figure 5). In some cases, the arene oxide has been identified as the actual enzymatic product that undergoes a nonenzymatic rearrangement to give the phenolic product with the concomitant migration and retention of substituent.[70] The formation of arene oxides from polycyclic aromatic hydrocarbons by these enzyme systems, in fact, caused a lot of public concern, as several of them are carcinogenic in nature.[71] The arene oxides derived from polycyclic aromatic hydrocarbons are more stable and isomerize very slowly to the corresponding phenols, compared to arene oxide derivatives of simple benzenoid compounds.

Extensive studies have been carried out on the mechanism of action of cytochrome P-450 monooxygenases,[67-69,72-75] a brief account of which is given below.

1. The substrate binds to the enzyme causing a distinct spectral change, indicating the formation of an ES complex (complex I) (Figure 9).
2. Cytochrome P-450 accepts two electrons in a sequential fashion from the iron-sulfur hemoprotein.
3. The addition of one electron leads to the ferrous form of the hemoprotein, which then binds to oxygen to form a complex that exists as either $Fe^{2+}O_2$ or $Fe^{3+}O_2^-$ (complex II).
4. If carbon monoxide is added instead of oxygen, an inactive enzyme-substrate-carbon monoxide complex is formed which exhibits an absorption maximum in the 450-nm region (complex III).
5. The cytochrome P-450-substrate-oxygen complex (complex II) undergoes one-electron reduction to give a $Fe^{3+}O_2^{2-}$ complex (complex IV). This complex may also arise by the interaction of the cytochrome P-450-substrate complex (complex I) with H_2O_2 as proposed by Coon et al.[52] Rapid reaction studies have provided evidence for the formation of both complex II and complex IV.

6. The uncoupling of hydroxylation caused by substrate analogs results in the dissociation of complex IV to give H_2O_2 and ES complex.
7. Complex IV loses a water molecule and is converted to an iron-oxygen complex, which is depicted as complex V. The oxygen in this complex is the activated species involved in the hydroxylation reaction. This complex has an overall charge of 3^+ and its formal oxidation state is 5^+.
8. Complex V hydroxylates the substrates to give the products. Complex V can also be formed by the interaction of the ES complex with organic peroxides and oxidizing agents such as iodosobenzene, sodium chlorite, and sodium periodate. At least in the case of iodosobenzene,[76] the pathway shown by Coon et al.[73] may not be operating and that proposed by Hrycay et al.[77] may be operating.

The catalytic mechanism of cytochrome P-450 monooxygenases is shown in Figure 9. This mechanism has extensive experimental support.

Comparatively little is known about plant cytochrome P-450 systems. One of the key enzymes involved in the hydroxylation of aromatic compounds in plants is the cinnamate-4-hydroxylase. This enzyme has been isolated from *Sorghum* and has clearly been shown to be a cytochrome P-450 monooxygenase by Conn's group.[78] The formation of 3,4-epoxide as the intermediate has also been established.[79] Recently, Lynen and group[80] have identified two hydroxylases, one converting *m*-cresol to 3-hydroxybenzylalcohol and toluquinol and the other oxidizing 3-hydroxybenzylalcohol to gentisylalcohol.[81] Both require NADPH and oxygen and seem to be cytochrome P-450 monooxygenases. In our laboratory, a phenyl-acetate-2-hydroxylase, which requires NADPH and oxygen, has been isolated and characterized to be a cytochrome P-450 monooxygenase.[82]

In view of the similarities observed by Smith et al.,[83-85] between mammalian hydroxylation and fungal hydroxylation of a number of aromatic compounds, it is likely that fungal systems possess a vast array of cytochrome P-450 type oxygenases. This field obviously requires more attention and offers itself as a fertile area for further studies.

d. Ascorbate-Linked Copper Proteins

Dopamine-β-hydroxylase is the only known example of this category. It hydroxylates dopamine (3,4-dihydroxyphenylethylamine) to norepinephrine, utilizing ascorbic acid and molecular oxygen.[69] It is a copper-containing monooxygenase. However, since it is a side-chain-hydroxylating enzyme, and is involved in metabolism of phenolic compounds, a short account ot its mechanism will be given here.

Kinetic studies[86] revealed the operation of the mechanism shown in Figure 10 for the hydroxylation reaction, which was further substantiated by electron paramagnetic resonance (EPR) studies.[87] EPR studies indicated that the copper atoms present in the enzyme are reduced by ascorbate to give Cu^+-enzyme. This form of the enzyme reacts with oxygen and substrate to give Cu^{2+}-enzyme, product, and water. Experiments with substrate stereospecifically labeled with tritium revealed that the reaction proceeds with the overall retention of configuration.[88]

e. Monophenol Monooxygenase Requiring Another Substrate Molecule as Reductant

Phenol-*o*-monooxygenase, variously termed phenolase, tyrosinase, or *o*-diphenol oxygen phenol oxidase, catalyzes the conversion of monophenols to *o*-diphenols, which are further oxidized to the corresponding *o*-quiones. A variety of phenol monooxygenases are found in bacteria, fungi, plants, insects, marine animals, and mammals, where they participate in the biosynthesis of lignins, flavonoids, tannins, adrenalin, and melanin. It is a copper-containing enzyme utilizing *o*-diphenol as the electron donor for the hydroxylation reaction.[69]

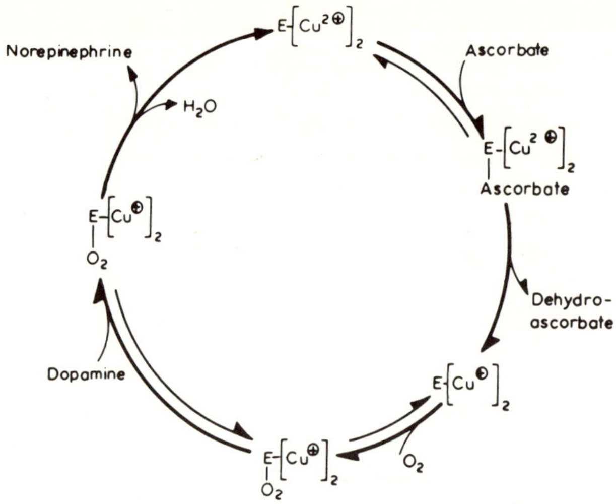

FIGURE 10. Sequence of reactions catalyzed by dopamine-β-hydroxylase.

2. Biosynthesis of Phenols by the Action of Dioxygenases

a. 4-Hydroxyphenylpyruvate Dioxygenases

4-Hydroxyphenylpyruvate dioxygenase catalyzes the simultaneous hydroxylation of the phenyl ring, migration of aliphatic side-chain to the adjacent position, and oxidative decarboxylation of the pyruvate fragment. Experiments with $^{18}O_2$ clearly indicated the dioxygenase nature of the enzyme.[89] Recent studies by Rundgren and group have established the presence of iron in this enzyme.[90,91] The enzyme occurs widely in mammalian systems. Two mechanisms have been proposed for the reaction catalyzed by this enzyme. One involves a quinone intermediate, while the other involves an epoxide intermediate. They are shown in Figure 11.

Experiments with chemically synthesized quinone intermediate and the purified enzyme revealed that the quinone is not a free intermediate in the enzyme-catalyzed reaction.[92] Steady state kinetic analysis revealed the probable operation of an ordered bi bi mechanism in which 4-hydroxyphenylpyruvate is bound first followed by oxygen and CO_2 is released prior to homogentisate.[93]

In addition to 4-hydroxyphenylpyruvate and phenylpyruvate, this enzyme also attacks 3,4-dihydroxyphenylpyruvate to give 2,4,5-trihydroxyphenylacetic acid.[94]

b. Hydroxylation of Aromatic Compounds by Dioxygenases

Recently, a group of new oxygenases have been discovered, which catalyze the double hydroxylation of a number of aromatic compounds. They seem to be present exclusively in prokaryotes. A number of aromatic compounds ranging from benzene to benzopyrene and benzoanthracene are oxidized by procaryotic organisms to their corresponding catechol derivatives, viz. *cis*-dihydrodiols.[95] The dioxygenase nature of such an enzyme system was first shown by Gibson's group[96] using $^{18}O_2$. Experiments with isotopic oxygen showed that both atoms of oxygen are incorporated into the product, viz. *cis*-1,2-dihydroxycyclohexa-3,5-diene. Subsequently, benzene dioxygenase was purified from a *Pseudomonas* species and was shown to consist of three protein components: A_1, A_2, and B.[97] A_1 was shown to be an iron-sulfur protein with a pair of antiferromagnetically coupled Fe^{3+} atoms. Component A_2 was found to be a flavoprotein, and B was shown to be a low M_r, relatively stable nonheme iron protein. All the components were necessary for the enzyme activity and absence of any one of them resulted in total loss of activity. In addition, both NADH and molecular

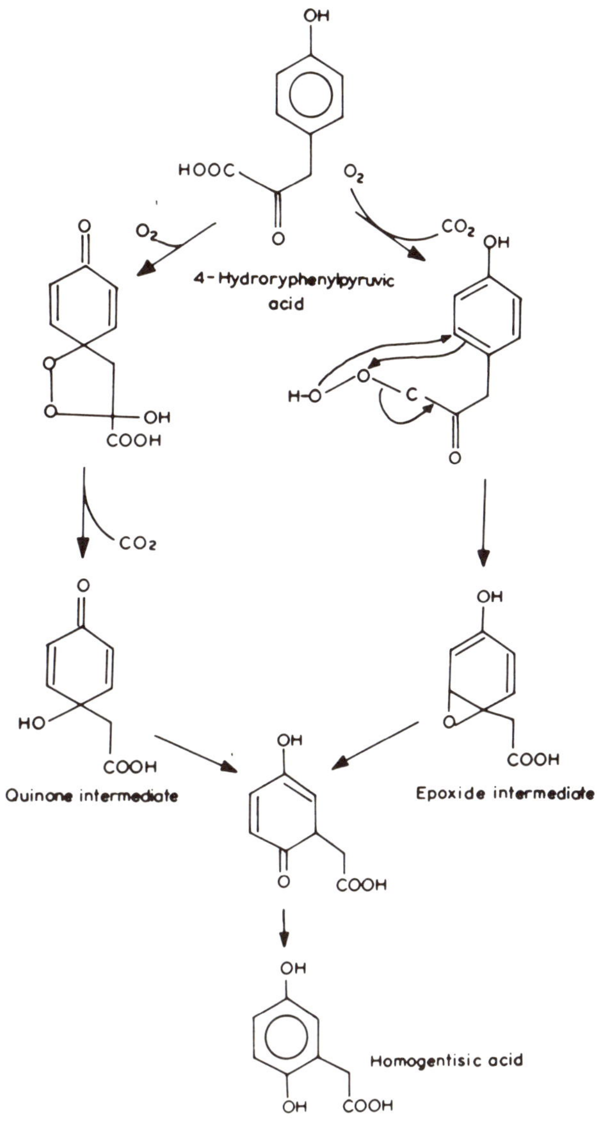

FIGURE 11. Mechanisms of action of 4-hydroxyphenylpyruvate dioxygenase.

oxygen were essential for oxygenation. It was suggested that organization of the components and cofactors is necessary for catalysis. The electrons from NADH are transferred to component A_2 with the concomitant reduction of bound FAD. The electrons are then transferred from component A_2 to component B and finally to component A_1. These transfers are mediated by the shuttling of the bound nonheme iron in the components A and B between Fe^{3+} and Fe^{2+} states. The electrons are finally transferred to benzene, which is converted to *cis*-1,2-dihydroxycyclohexa-3,5-diene with the incorporation of both atoms of molecular oxygen. The *cis*-1,2-dihydroxycyclohexa-3,5-diene thus formed is further converted to catechol by a separate dehydrogenase which requires NAD as the cofactor.[98] The conversion of benzene to catechol and the functions of the components of benzene dioxygenase are shown in Figure 12.

Similar enzyme systems, which convert toluene into 3-methylcatechol, have also been characterized.[99,100] Similarly, naphthalene, phenanthrene, anthracene, xylenes, biphenyl,

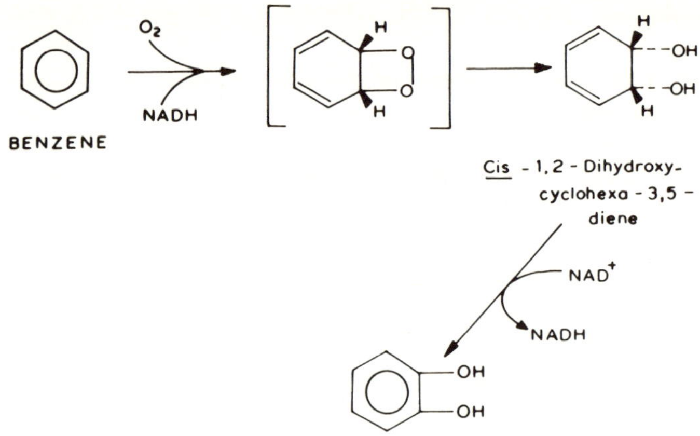

FIGURE 12(a). Mechanism for the conversion of benzene to catechol.

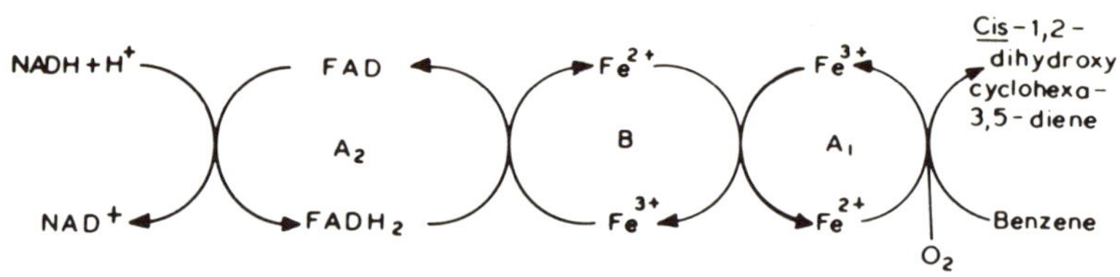

FIGURE 12(b). Components of benzene dioxygenase.

benzopyrene, and benzoanthracene are converted to their corresponding catechol derivatives.[95] Apart from aromatic hydrocarbons, benzoic acid and aniline derivatives are also converted to the catechol derivatives by dioxygenase enzyme systems.[101] Using a mutant strain of *Alcaligenes eutrophus,* Reiner and Hegeman[102] showed the formation of 1,2-dihydroxycyclohexa-3,5-diene-1-carboxylic acid from benzoic acid. These workers also proved,with the help of $^{18}O_2$ studies, that the enzyme system involved is a dioxygenase.

Yamaguchi et al.[103,104] purified benzoate-1,2-dioxygenase from the cell-free extracts of *P. arvilla* and showed that this enzyme consists of two protein components, A and B. Component A was found to possess very high NADH-cytochrome *c* reductase activity, while component B was identified to be the oxygenase. Reducing equivalents from NADH were transferred to the oxygenase via the reductase. The reduced oxygenase then oxygenates benzoate to 1,2-dihydroxycyclohexa-3,5-diene-1-carboxylic acid. Thus, the benzoate dioxygenase system described by Yamaguchi et al.[103,104] seems to be a much simpler system when compared to the aromatic hydrocarbon dioxygenases which are known to be three-component enzyme systems.[97,99] The 1,2-dihydroxycyclohexa-3,5-diene-1-carboxylic acid thus formed undergoes both dehydrogenation and decarboxylation to give catechol by a separate enzyme system.[105] This enzyme system, which has been purified to homogeneity from *Alcaligenes eutrophus,* has a M_r of about 94,600 and specifically requires NAD as the cofactor.

p-Toluic acid, as well as *m*-toluic acid, seems to undergo this type of oxidative decarboxylation.[106] Both cumic acid[107,108] and *o*-phthalic acid[109] undergo oxygenation reactions without decarboxylation. The dihydrodiols in both cases have been characterized. *o*-Phthalate dioxygenase has been purified and shown to be a two-component enzyme system[109] similar to the benzoate dioxygenase.[103]

Both aniline[110] and anthranilic acid[111] are converted to catechol by procaryotic organisms, again employing a dioxygenation reaction as evidenced by $^{18}O_2$ studies. Anthranilate dioxygenase from *Pseudomonas* requires NADH and molecular oxygen and is made up of two protein components.[112]

The enzymology of the reactions mentioned above has not been fully studied and requires a more detailed investigation.

3. Biosynthesis from Arene Oxides by Hydration, Dehydrogenation Reactions

Arene oxide hydration to yield *trans*-dihydrodiols that can undergo dehydrogenation to give *o*-dihydroxy compounds is a minor route for their biosynthesis. Nevertheless, the enzyme systems responsible for arene oxide hydration seem to play a crucial role in determining the biological fate of many environmental pollutants and toxic compounds. These are converted to arene oxides by cytochrome P-450 monooxygenases that are of widespread occurrence in nature. Hence, it is logical to expect the presence of epoxide hydrases in several systems.

As early as 1950, Boyland[113] suggested that *trans*-dihydrodiols might arise from arene oxides. Subsequently, Boyland and Sims provided evidence for the conversion of epoxides to dihydrodiols.[114,115] Later, Jerina et al.[116] carried out a detailed study on the epoxide hydration reaction and characterized the enzyme system[117] involved. The enzyme which occurs in the membrane fraction has a broad substrate specificity and hydrates a number of epoxides and arene oxides. Lu et al.[118,119] have purified epoxide hydrase and studied its kinetic properties with a number of epoxides and arene oxides. Whether or not all the dihydrodiols formed in this way undergo dehydrogenation to give catechol derivatives awaits confirmation. However, Ayengar et al.[120] demonstrated the presence of an NADP-dependent enzyme in rabbit liver which causes the dehydrogenation of 3,5-cyclohexadiene-1,2-diol to catechol.

In the fungus *Cunninghamella elegans,* naphthalene seems to undergo such a reaction sequence, whereby it is converted to 1,2-dihydroxynaphthalene.[121,122]

III. DEGRADATION OF PHENOLS

The large number of phenolic compounds formed by the operation of innumerable pathways, discussed in the previous section, along with several pollutants, enter the soil and have to be finally degraded by the action of dioxygenases. Dioxygenases act only on di- and polyhydric phenols and not on monohydric phenols. In the following section, a survey of dioxygenases and their possible mechanism of action will be discussed.

A. Survey of Dioxygenases

1. Flavonoid Dioxygenases

An interesting dioxygenase characterized by Simpson's group oxidizes quercetin to 2-protocatechuoyl phloroglucinol-carboxylic acid. Using $^{18}O_2$ it was shown that the degradation of rutin in *Aspergillus flavus* was catalyzed by an extracellular dioxygenase of M_r 110,000.[123,124] It is a unique glycoprotein oxygenase with copper as the prosthetic group.[125] This enzyme is made up of two subunits, each having one substrate-binding site and a g-atom of copper.[126] Jeffrey et al.,[127] who studied the degradation of dihydroquercetin by a soil *Pseudomonas* grown on catechin, have reported an oxygenase which cleaves dihydrogossypetin by an extradiol mechanism. Wood's group[128] have also characterized a new oxygenase involved in the degradation of quercetin by *P. putida.*

2. Ring Cleavage of Quinols

Scott and Beadling[129] have characterized a dioxygenase catalyzing the ring cleavage of gentisaldehyde to yield 3-formyl-4-hydroxymuconic semialdehyde, a step in the biosynthesis of patulin. The enzyme catalyzing the ring cleavage of the corresponding acid, namely gentisic acid, has been purified from a *Pseudomonas* species by Lack[130] and by Sugiyama et al.[131] It requires Fe^{2+} for activity.

Recently, Crawford et al.[132] have isolated this enzyme from *Moraxella osloensis* and purified it to homogeneity. The homogeneous enzyme appears to be a tetramer made up of identical subunits of M_r 40,000. In addition to gentisate, a number of substituted gentisic acids serve as substrates for this enzyme, but with less efficiency.

The dioxygenase which attacks homogentisic acid is the most widely distributed quinol dioxygenase. Its presence in several microbes like *Pseudomonas,*[133] *Vibrio,*[134] *Bacillus,*[135] *Moraxella,*[135] *Aspergillus niger,*[136] and plant[137] and mammalian systems[138-141] has been reported. The purified enzyme requires specifically Fe^{2+} and sulfhydryl groups for activity. The fungal[136] and bacterial[133] enzymes have been purified to homogeneity. The bacterial enzyme has been crystallized. The enzyme from prokaryotes has a M_r of 380,000, while that of the eucaryotic enzyme has a M_r of approximately 200,000.[133,136,141]

3. 3-Hydroxyanthranilate Dioxygenase

Although 3-hydroxyanthranilate dioxygenase has been characterized from a number of mammalian systems, it could not be purified to homogeneity for quite some time due to its extreme instability. Recently, however, Koontz and Shiman[142] purified this enzyme to homogeneity. It is a single polypeptide of M_r 34,000 and readily oxygenates to form inactive, higher M_r oligomers. This enzyme also requires Fe^{2+} for its activity.

4. Ring Cleavage of DOPA

The cleavage of the ring of DOPA seems to occur in higher plants as well as in microorganisms. The brightly colored betalin pigments which are widespread in *Centrospermae* arose from DOPA by extradiol ring fission followed by ring closure reactions.[143,144]

Stizolobinic acid and stizolobic acid, a new type of unusual, heterocyclic amino acids, found in five genera of *Lotoideae* and in some species of toxic mushrooms belonging to the Basidiomycetes family, arise from DOPA by an extradiol cleavage reaction and further cyclization. The enzymes responsible for the synthesis of these two amino acids have been purified from cell-free extracts of the higher plant *Stizolobium hasjoo* and shown to require NAD^+P.[145]

Similarly, muscaflavin seems to arise from DOPA by extradiol cleavage followed by cyclization.[146] Experiments using labeled compounds indicate that during the biosynthesis of the antitumor antibiotic, demethyltomamycin, in *Streptomyces achromogenes,* DOPA ring cleavage might occur.[147] These reactions are summarized in Figure 13.

5. Cleavage of ortho-Dihydroxy Phenolic Compounds

P. fluorescens grown on *p*-coumaric acid was able to catalyze the cleave of the ring of caffeic acid by an intradiol mechanism, i.e., bond breakage between the vicinal hydroxyl groups.[148] Dagley et al.[149] observed the ring cleavage of 2,3-dihydroxyphenylpropionate between 1- and 2-positions by a dioxygenase isolated from *Achromobacter.*

In the case of homoprotocatechuic acid and protocatechuic acid, all three possible modes of ring cleavage are known. *Pseudomonas ovalis,*[150] *Bacillus stearothermophilus,*[151] and *Arthrobacter,*[152] possess a 2,3-dioxygenase, while the *Pseudomonas* species isolated by Dagley and Wood[153] elaborates a 4,5-dioxygenase for the utilization of homoprotocatechuate. Both of these enzymes are nonheme dioxygenases. The crystalline 2,3-dioxygenase has a M_r of about 140,000 and seems to have four identical subunits of M_r 35,000 which contain approximately 1 g-atom of ferric iron.[154-157] Subba Rao et al.[158] have isolated an enzyme from the fungus *Tilletiopsis washingtonensis,* which cleaves homoprotocatechuic acid by an intradiol mechanism. Surprisingly, there is no evidence for any metal ion requirement for this enzyme system.

Recently it has been shown that homoprotecatechuate 2,3-dioxygenase from *Bacillus brevis* is a manganese (II)-dependent enzyme. It is not inhibited by cyanide, DDC, and hydrogen

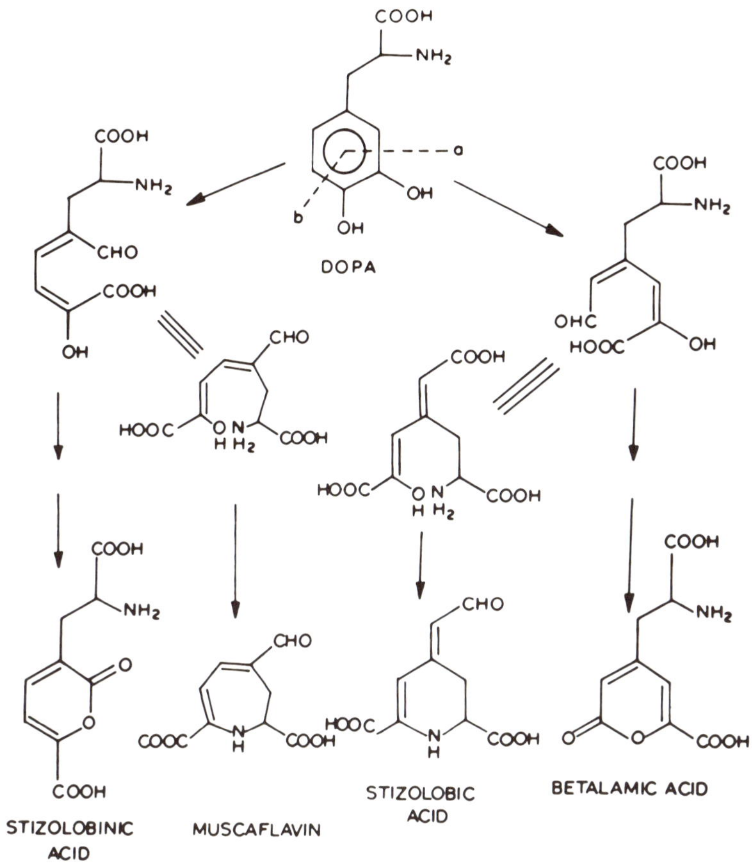

FIGURE 13. Ring cleavage of DOPA.

peroxide (H_2O_2).[159] The enzyme is similar to *Arthrobacter*[152] and *Pseudomonas*[153] enzyme in its molecular weight and subunit structure. X-ray, fluorescence, and atomic absorption spectroscopy showed the presence of 2 g-atoms of manganese per mole of protein.

Protocatechuate 4,5-dioxygenase was the first *meta*-cleaving (extradiol) dioxygenase to be characterized.[160] Detailed studies on this enzyme have been carried out by Ono et al.,[161] as well as by Zabinski et al.[162] While both groups of workers reported a M_r of about 15,000 for this enzyme, they estimated the iron content as 1 and 4 g-atoms per mole enzyme, respectively.

Protocatechuate 2,3-dioxygenase is present in *B. circulans.*[163] Protocatechuate-3,4-dioxygenase, however, is present in a number of microorganisms, including an unidentified soil bacterium,[164] *Pseudomonas* species,[165-167] *Acinetobacter calcoaceticus,*[168] *Nocardia,*[169] *Neurospora,*[170] *and Tilletiotopsis washingtonensis.*[158] The enzymes from the *P. aeruginosa*[167] and *Acinetobacter*[168] have been purified to homogeneity and those from the former source crystallized;[171] the α-subunit has been sequenced.[172,173] While these two enzymes seem to have a M_r of about 700,000, that from *P. putida* has a M_r of only 200,000.[166] The dioxygenase from *P. aeruginosa* has eight subunits, each having a substrate binding site and 1 g-atom of iron.[174] Each subunit in turn seems to consist of four smaller subunits of two different types — α_2 β_2.[175]

Pyrocatechuate (2,3-dihydroxybenzoate) ring cleavage was first reported by Ribbons[176] in *P. fluorescens.* Subsequently, Madhyastha et al.[177] reported pyrocatechuate ring cleavage between 1,2-positions by a soil *Pseudomonas* grown on *p*-cymene. However, DeFrank and

Ribbons[178] carefully studied the reactions and concluded that the ring cleavage occurs between 3- and 4-positions.

From our laboratory, we have reported the intradiol cleavage of pyrocatechuic acid in the higher plant *Tecoma stans*.[179] Interestingly, this enzyme seems to be a copper protein similar to the quercetinase.[125]

Catechol dioxygenases are relatively less specific enzymes and attack a number of catechol derivatives. Pyrocatechase was the first enzyme shown to be dioxygenase by $^{18}O_2$ studies.[180] It has been purified from *P. arvilla*,[181] *P. fluorescens*,[182] *Acinetobacter calcoaceticus*,[183] and *Brevibacterium fuscum*.[184] The M_r of catechol 1,2-dioxygenase ranges from 64,000 to 90,000. Fujiwara et al. have shown that the *Pseudomonas* enzyme catalyzes not only an intradiol cleavage, but also an extradiol cleavage when 3-methylcatechol is used as the substrate.[185] Quite recently, Que[186] has shown that pyrocatechase oxidizes *o*-aminophenol by an extradiol cleavage mechanism to an extent of 95%.

meta-Pyrocatechase, also known as catechol 2,3-dioxygenase, was the first dioxygenase to be obtained in a crystalline form.[187,188] It has a M_r of 140,000 and seems to be made up of 3 or 4 subunits, each having 1 g-atom of iron.

4-Methylcatechol undergoes all three possible modes of cleavage. For instance, *P. desmolyticum* oxidizes 4-methylcatechol by an extradiol proximal as well as an intradiol cleavage mechanism,[189] while a yeast strain utilizes it by an extradiol distal cleavage mechanism.[190,191] Several strains of bacteria oxidize methylcatechols by extradiol proximal cleavage, while a few species cleave them by an extradiol mechanism.[192]

6. Dioxygenases Acting on Trihydroxy Compounds

Although two hydroxyl groups or one amino and one hydroxyl group are sufficient to labilize the aromatic ring, some dihydroxy compounds undergo a third hydroxylation to form trihydroxyphenols before they are cleaved. Thus, orcinol, resorcinol, quinol, and thymol are converted to the trihydroxy compounds before the ring is cleaved. Gallic acid undergoes ring cleavage to give 5-hydroxy-3-carboxy-*cis, cis*-muconic acid in *P. putida*.[198] Orcinol as well as resorcinol are further hydroxylated to trihydroxy derivatives and then cleaved by *P. putida* to give 2,4,6-trioxoheptanoic acid and 2,4-dihydroxymuconic semialdehyde, respectively.[194] The same organism under certain growth conditions cleaves resorcinol by an intradiol cleavage mechanism, after converting it into hydroxyquinol. The formation of maleylacetic acid has also been demonstrated in the case of quinol.[195] Thymol also follows this pattern and is converted to hydroxythymoquinone before its ring is cleaved in *P. putida*.[196] Dagley's group[197] have also shown the further hydroxylation of 3,4-dihydroxyphenylacetic acid to give 2,4,5-trihydroxyphenylacetic acid before the ring is cleaved by *Trichosporon cutaneum*.

B. Mechanism of Action of Dioxygenases

There are several excellent review articles available on the various modes of cleavage of catechols and quinol derivatives as well as on the oxygenases involved in these pathways.[11,191,198-206] However, the mechanism of cleavage of phenol ring systems has not been discussed in depth in any of the review articles, with the exception of the paper by Wood et al.,[205] who have given a brief account of the mechanism of oxygenases. This part of the review will deal with the mechanistic features of dioxygenases. All substrates for ring-cleaving dioxygenases are singlet molecules, while cosubstrate, i.e., oxygen, is a triplet molecule in the ground state. Hence, as pointed out by Hamilton,[207] the reaction of singlet molecules with triplet oxygen to give singlet products is a spin-forbidden process and will not occur readily. This necessitates the activaiton of either molecular oxygen or the substrate. The available evidence suggests that both possibilities might take place in biological systems.

1. Oxygen Activation

a. Singlet Oxygen Involvement

The singlet species is often implicated as an active oxygen intermediate in biological processes. This intermediate rapidly reacts with olefins to give dioxetanes, which ultimately rearrange to form two carbonyl fragments. Thus, the reaction of singlet oxygen with indene gives a dioxetane intermediate, which is rapidly decomposed to give homophthalaldehyde.[208]

However, there is no evidence for the participation of singlet oxygen in any ring-cleaving system. First, it is not clear how an enzyme could supply energy of about 22kcal/mol to excite the triplet oxygen to the singlet level. Second, the reaction of singlet oxygen has very little resemblance to enzymic dioxygenations. Again, dioxetane intermediates, on decomposition, emit light.[208] Thus far, there is no evidence of chemiluminescence in any reaction resulting in ring cleavage. Most reactions of singlet oxygen with organic compounds give peroxide products. Only a few oxygenase-catalyzed reactions like lipoxygenase reaction give peroxide as the ultimate product. Hence, it is unlikely that singlet oxygen is involved in dioxygenase reactions.

b. Superoxide Anion Involvement

Superoxide anion (O_2^-), the univalently reduced species of molecular oxygen, has received considerable attention during the past few years, as an active species of oxygen in certain oxygenase reactions.

During their investigation on the chemistry of superoxide anion, Moro-Oka and Foote[209] reported the ring cleavage of 3,5-ditertiarybutylcatechol as well as 9,10-dihydroxyphenanthrene with superoxide anions. Essentially similar results were also reported by Lee-Ruff et al.[210] In the same year, Nishinaga et al.[211] reported base-catalyzed oxidative ring fission of 3,5-ditertiarybutylcatechol and its *ortho*-benzoquinone. All of these workers proposed the formation of semiquinone as an obligatory intermediate in ring fission.

Although the evidence for the participation of O_2^- in aromatic ring cleavage reactions is not completely convincing, clear-cut evidence exists for its involvement in the cleavage of the ring of tryptophan catalyzed by indoleamine-2,3-dioxygenase.[212,213] These studies were further confirmed using potassium superoxide as the substrate for the dioxygenase reaction.[214] Hayaishi's group[215] also characterized and studied the properties of enzyme-superoxide anion complex using fast reaction kinetics. Studies with $K^{18}O_2^-$ and $^{18}O_2$ revealed that 60% of the oxygen atoms incorporated into the product came from $K^{18}O_2^-$ and about 40% from molecular oxygen, suggesting that the enzyme preferentially utilizes superoxide anion as the substate for the ring-cleaving reaction.[216]

From our laboratory, the possible participation of superoxide anion in the homogentisate dioxygenase-catalyzed reaction by inhibition studies with superoxide dismutase has been reported.[136]

C. Peroxide Involvement

Cleavage of the aromatic ring by hydrogen peroxide as well as by peracids has been known for some time. Thus, as early as 1931, Boeseken and Engelberts[217] reported the oxidative cleavage of phenol to *cis, cis*-muconic acid using peracetic acid. Recently, Pandell[218] has reported metal-catalyzed peracetic acid oxidation of phenol and catechol to *cis, cis*-muconic acid. However, in all of these studies it was observed that 2 mol of peracetic acid are required for the oxidation of each mole of catechol, while in the oxygenase-catalyzed reactions, only 1 mole of oxygen is required.

The reaction of 3,5-ditertiarybutyl-*o*-benzoquinone with H_2O_2 is known to proceed under alkaline conditions to give ring-cleaved products. However, the similarity of this system to enzymatic reactions is not appreciable.

2. Substrate Activation

Till recently there was very little evidence for substrate activation during any dioxygenase reaction. However, preliminary studies with model systems suggested the operation of such a mechanism. Since oxygen activation is generally associated with metal ions, results with metal ion-free model systems first gave a clue for the possible activation of the substrate. For instance, catechols react rapidly with molecular oxygen in the absence of any transition metal ion, under alkaline conditions,[211,219] to give ring-cleaved products. Baldwin[220] observed that 3,6-ditertiary-4-*O*-methylcatechol undergoes rapid isomerization to the keto-alcohol form at −50°C in the absence of oxygen. Upon exposure to oxygen, the corresponding *cis, cis*-muconic acid was formed. These experiments suggest that oxygen can indeed add to organic stubstrates and form perhaps a peroxy intermediate, which might undergo ring cleavage to give *cis, cis*-muconic acid. This type of substrate activation seems to occur in the protocatechuate-3,4-dioxygenase reaction.[205,221]

3. Mechanism

Any mechanism for ring cleavage should explain how the two atoms of oxygen end up on the carbon atoms that were adjacent to each other in the original substrate molecule. Initially, a four-membered cyclic dioxetane has been proposed as an intermediate, which is possibly formed from the peroxide precursor. The dioxetanes thus formed are highly unstable and decompose readily to form ring fission products. The proposed mechanism of cleavage of the catechol ring by bacterial dioxygenases is shown in Figure 14. However, as stated earlier, there is no chemiluminescence during dioxygenase action, and energy calculations show that the proposed mechanism is not possible.[207]

Hamilton[222] proposed an alternative mechanism, which is shown in Figure 15. The peroxy intermediate formed undergoes rearrangement to give a seven-membered cyclic intermediate. This rapidly opens up to give the ring-cleaved product. Both of the above mechanisms involve the activation of oxygen by the metal ion, while the mechanism proposed by Wood and group[205,221] invokes the activation of the substrate by the metal ion.

The iron atoms in protocatechuate dioxygenase serve to bind the substrate and promote the polarization of electrons toward C-4. Oxygen then reacts with the enzyme-bound polarized substrate molecule to form a peroxide type of intermediate. Iron also plays a crucial role in the formation of this peroxide intermediate, possibly by an ionic mechanism. The rest of the steps are the same as suggested by Hamilton[222] (Figure 15).

Mossbauer and EPR spectroscopic studies clearly established that the iron atoms in protocatechuate dioxygenase stay in high spin ferric form throughout the course of the reaction cycle,[221] which precludes the direct binding of oxygen to the ferric iron, as only high spin ferrous ion is known to bind oxygen. Kinetic data supported the fact that the 4-hydroxyl group binds to the enzyme-bound ferric iron. This causes the promotion of electron migration towards C-4, which then reacts with oxygen to form semiquinone and superoxide. The superoxide thus formed binds to iron and the two radicals eventually couple to give a peroxy intermediate that decomposes to give the ring-cleaved products. Whether or not all dioxygenases follow this mechanism remains to be established, although in the case of some of the dioxygenases, the iron atom oscillates between ferrous and ferric forms during catalysis.[199-203] Further studies are obviously necessary to throw more light on this question.

IV. FUTURE TRENDS AND SCOPE FOR FURTHER STUDIES

The review of literature presented in these pages clearly highlights the diversity in the pathways employed by microorganisms to metabolize phenolic compounds. Oxygenases occupy a pivotal position in the utilization of these compounds. A study of these enzymes has been extremely rewarding in the understanding of the mechanisms of oxygen activation

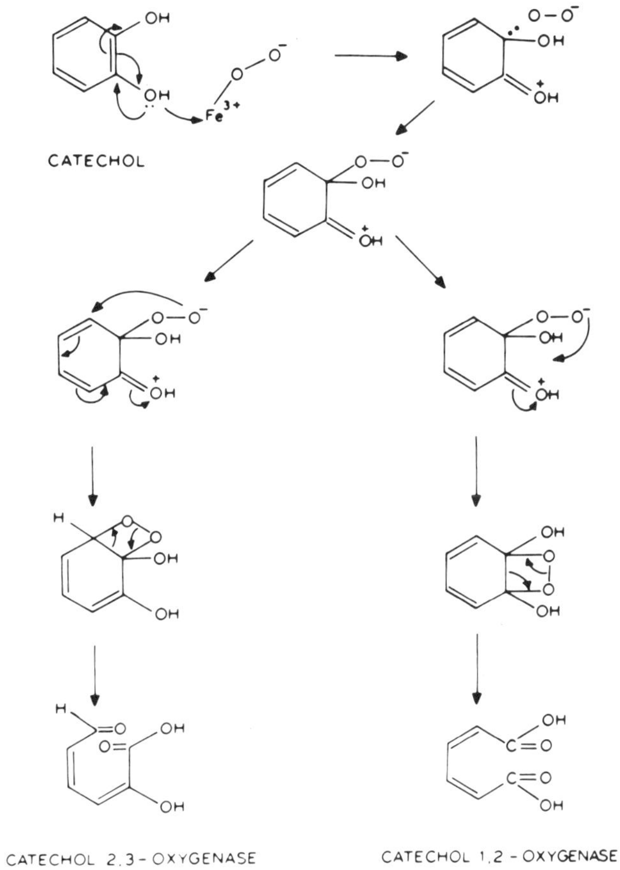

FIGURE 14. Proposed mechanisms for the cleavage of the catechol ring by bacterial dioxygenases.

and the role of flavins, pyridine nucleotides, pteridines, metal ions, cytochromes, and nonheme iron thereof.

The pathways of degradation of aromatic compounds and the enzymes involved have been studied mainly in bacterial species belonging to the genus *Pseudomonas*. Nevertheless, more studies in other bacterial systems are necessary in order to evaluate the universal and evolutionary significance of the catabolic pathways. Despite the large number of papers that appear on the mechanism of oxygenases, our knowledge on the active molecular oxygen species involved in these reactions remains obscure. Further studies are necessary to understand the role of various cofactors in the oxygenase reactions. Superoxide anions, hydroxy radicals, and peroxy radicals have all been implicated in oxygenase reactions. However, conclusive evidence for any of these intermediates is still meager. Recently, interest has been shown in studies on the degradation of recalcitrant compounds by microorganisms[223] and on the construction of hybrid organisms capable of degrading a variety of aromatic compounds.[224] In the future, more studies in this direction are necessary to combat environmental pollution.

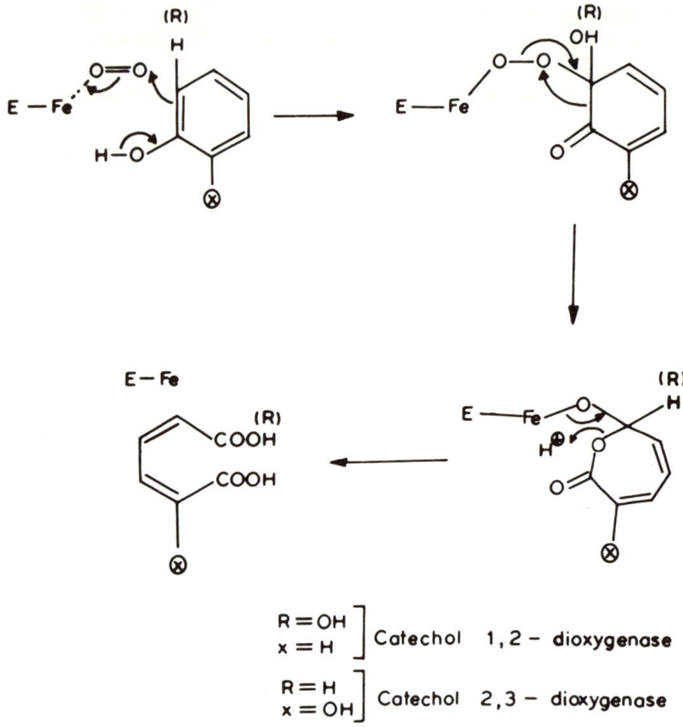

FIGURE 15. Proposed mechanisms for the cleavage of the catechol ring (Hamilton).

ACKNOWLEDGMENTS

The author wishes to express his deep sense of gratitude to Dr. (Mrs.) V. Geeta Rani, Mr. B. Haribabu, and Mr. Ajith V. Kamath for their valuable help and unstinted cooperation in the collection of literature and the preparation of the manuscript.

REFERENCES

1. **Haslam, E.,** *The Shikimate Pathway,* Butterworths, London, 1974.
2. **Richards, J. R. and Hendrickson, J. B.,** *The Biosynthesis of Steroids, Terpenes and Acetogenins,* W. A. Benjamin, New York, 1964, 27.
3. **Dimroth, P., Walter, H., and Lynen, F.,** Biosynthese von 6-Methylsalicylsaure, *Eur. J. Biochem.,* 13, 98, 1970.
4. **Dimroth, P., Ringelmann, E., and Lynen, F.,** 6-Methylsalicylic acid synthetase from *Pencillium patulum, Eur. J. Biochem.,* 68, 591, 1976.
5. **Lynen, F.,** in *Organizational Biosynthesis,* Vogel, H. J. et al. Eds., Academic Press, New York, 1967, 243.
6. **Cordell, G. A.,** *Chem. Rev.,* 76, 425, 1976.
7. **Schubert, K.,** *Biosynthesis of Aromatic Compounds,* Billek, G., Ed., Pergamon Press, Oxford, 1966, 81.
8. **Moss, G. P.,** in *Terpenoids and Steroids,* A Specialist Periodical Report, Vol.2, The Chemical Society, London, 1972, 197.
9. **Kaneda, T.,** Enzymatic aromatization of 4-ketocyclohexane carboxylic acid to p-hydroxybenzoic acid, *Biochem. Biophys. Res. Commun.,* 58, 140, 1974.
10. **Boyer, P. D., Ed.,** *The Enzymes,* Vol. 12, Academic Press, New York, 1975.
11. **Hayaishi, O., Ed.,** *Molecular Mechanisms of Oxygen Activation,* Academic Press, New York, 1974.

12. **Kaufman, S.,** *Adv. Enzymol.,* 35, 245, 1971.
13. **Kaufman, S. and Fisher, D. B.,** in *Molecular Mechanisms of Oxygen Activation,* Hayaishi, O., Ed., Academic Press, New York, 1974, 285.
14. **Guroff, G., Daly, J. W., Jerina, D. W., Renson, J., Witkop, B., and Udenfriend, S.,** Hydroxylation induced migration: the NIH shift, *Science,* 157, 1524, 1967.
15. **Fisher, D. B., Kirkwood, R., and Kaufman, S.,** Rat liver phenylalanine hydroxylase, an iron complex enzyme, *J. Biol. Chem.,* 247, 5161, 1972.
16. **Letendre, C. H., Dickens, G., and Guroff, G.,** Phenylalanine hydroxylase from *Pseudomonas* sp. (ATCC 11299a), *J. Biol. Chem.,* 250, 6672, 1972.
17. **Nakata, H., Yamaguchi, T., and Fujisawa, H.,** Phenylalanine hydroxylase from *Chromobacterium violaceium, J. Biol. Chem.,* 254, 1829, 1979.
18. **Shiman, R. and Gray, D. W.,** Substrate activation of phenylalanine hydroxylase, *J. Biol. Chem.,* 255, 4793, 1980.
19. **Kaufman, S. and Mason, K.,** *J. Biol. Chem.,* 257, 14,667, 1982.
20. **Shiman, R., Mortimore, G. E., Schworer, C. M., and Gray, D. W.,** *J. Biol. Chem.,* 257, 11213, 1982.
21. **Shiman, R. and Jefferson, L. S.,** Iron dependent regulation of rat phenylalanine hydroxylase activity *in vivo* and *in vitro* and in perfused liver, *J. Biol. Chem.,* 257, 839, 1982.
22. **Donlon, J. and Kaufman, S.,** Relationship between the multiforms of rat hepatic phenylalanine hydroxylase and degree of phosphorylation, *J. Biol. Chem.,* 255, 2146, 1980.
23. **Parniak, M. A. and Kaufman, S.,** Rat liver phenylalanine hydroxylase activation by sulfhydryl modification, *J. Biol. Chem.,* 256, 6876, 1981.
24. **Kaufman, S.,** The activity of 2,4,5-triamino-6-hydroxy pyrimidine in the phenylalanine hydroxylase system, *J. Biol. Chem.,* 254, 5150, 1979.
25. **Baily, S. W., Weintraub, S. T., Hamilton, S. M., and Ayling, J. E.,** Incorporation of molecular oxygen into pyrimidine cofactors by phenylalanine hydroxylase, *J. Biol. Chem.,* 257, 8253, 1982.
26. **Hoeldtke, R. and Kaufman, S.,** Bovine adrenal tyrosine hydroxylase purification and properties, *J. Biol Chem.,* 252, 3160, 1977.
27. **Letendre, C. H., Dickens, G., and Guroff, G.,** The tryptophan hydroxylase of *Chromobacterium violaceium, J. Biol. Chem.,* 249, 7186, 1974.
28. **Chenna Reddy, C. and Vaidyanathan, C. S.,** Purification and properties of benzoate-4-hydroxylase from a soil *Pseudomonas, Arch. Biochem. Biophys.,* 177, 488, 1976.
29. **Chenna Reddy, C. and Vaidyanathan, C. S.,** *Biochim. Biophys. Acta,* 384, 46, 1975.
30. **Bhat, S. G. and Vaidyanathan, C. S.,** Purification and properties of L-mandelate-4-hydroxylase from *Pseudomonas convexa, Arch. Biochem. Biophys.,* 176, 314, 1976.
31. **Kohmoto, K. and Nishimura, S.,** Phenylacetic acid *meta* hydroxylase from *Rhizoctonia solani, Phytochemistry,* 14, 2131, 1975.
32. **Flashner, M. S. and Massey, V.,** *Molecular Mechanisms of Oxygen Activation,* Hayaishi, O., Ed., Academic Press, New York, 1974, 245.
33. **Massey, V. and Hemmerich, P.,** in *The Enzymes,* Vol. 12, Boyer, P.D., Ed., Academic Press, New York, 1975, 191.
34. **Massey, V.,** *Biochemical and Clinical Aspects of Oxygen,* Caughey, W. S., Ed., Academic Press, New York, 1979, 477.
35. **Takemori, S., Nakamura, M., Suzuki, K., Katagiri, M., and Nakamura, T.,** *Biochim. Biophys. Acta,* 247, 2358, 1972.
36. **White-Stevens, R. H. and Kamin, H.,** Studies of a flavoprotein: salicylate hydroxylase, *J. Biol. Chem.,* 247, 2358, 1972.
37. **Entsch, B., Ballou, D. P., and Massey, V.,** Flavin-oxygen derivatives involved in hydroxylation by *p*-hydroxybenzoate hydroxylase, *J. Biol. Chem.,* 251, 2550, 1976.
38. **Ghisla, S., Entsch, B., Massey, V., and Husein, M.,** On the structure of flavin-oxygen intermediate involved in enzymatic reactions, *Eur. J. Biochem.,* 76, 139, 1977.
39. **Entsch, B., Ballou, D. P., Husein, M., and Massey, V.,** Catalytic mechanism of p-hydroxybenzoate hydroxylase with *p*-mercaptobenzoate as substrate, *J. Biol. Chem.,* 251, 7367, 1976.
40. **Schopfer, L. M., Massey, V., and Claiborne, A.,** Active site probes of flavoproteins, *J. Biol. Chem.,* 256, 7329, 1981.
41. **Massey, V., Ghisla, S., and Moore, E. G.,** *J. Biol. Chem.,* 254, 9640, 1979.
42. **Moore, E. G., Ghisla, S., and Massey, V.,** Properties of flavins where the 8-methyl group is replaced by mercapto residues, *J. Biol. Chem.,* 254, 8173, 1979.
43. **Claiborne, A., Massey, V., Fitzpatrick, P. F., and Schopfer, L. M.,** 2-Thio flavins as active site probes of flavoproteins, *J. Biol. Chem.,* 257, 174, 1982.
44. **Entsch, B., Husein, M., Ballou, D. P., Massey, V., and Walsh, C.,** Oxygen reactivity of p-hydroxybenzoate hydroxylases containing 1-deaza FAD, *J. Biol. Chem.,* 255, 1420, 1980.

45. **Claiborne, A., Hemmerich, P., Massey, V., and Lawton, R.,** *J. Biol. Chem.*, 258, 5433, 1983.
46. **Biemann, M., Claiborne, A., Ghisla, S., Massey, V., and Hemmerich, P.,** *J. Biol. Chem.*, 258, 5440, 1983.
47. **Shoun, H., Beppu, T., and Arima, K.,** On the stable enzyme substrate complex of *p*-hydroxybenzoate hydroxylase, *J. Biol. Chem.*, 254, 899, 1979.
48. **Shoun, H., Higashi, N., Beppu, T., Nakamura, S., Hiromi, K., and Arima, K.,** Studies on the interaction of *p*-hydroxybenzoate hydroxylase with NADPH, *J. Biol. Chem.*, 254, 10,944, 1979.
49. **Shoun, H., Beppu, T., and Arima, K.,** An essential arginine residue at the substrate binding site of *p*-hydroxybenzoate hydroxylase, *J. Biol. Chem.*, 255, 9319, 1980.
50. **Shoun, H. and Beppu, T.,** *J. Biol. Chem.*, 257, 3422, 1982.
51. **Levy, C. C.,** Melilotate hydroxylase, *J. Biol. Chem.*, 242, 747, 1968.
52. **Strickland, S. and Massey, V.,** The purification and properties of melilotate hydroxylase, *J. Biol. Chem.*, 248, 2944, 1973.
53. **Strickland, S., Schopfer, L., and Massey, V.,** Kinetic and mechanistic studies on the reaction of melilotate hydroxylase with deuterated melilotate, *Biochemistry*, 14, 2230, 1975.
54. **Daly, J. W., Guroff, G., Udenfriend, S., and Witkop, B.,** Hydroxylation-induced migration of tritium in several substrates of liver aryl hydroxylases, *Arch. Biochem. Biophys.*, 122, 218, 1967.
55. **Ohta, T., Higgins, D. W., and Ribbons, D. W.,** Metabolism of resorcinylic compounds by bacteria — purification and properties of orcinol hydroxylase from *Pseudomonos putida, J. Biol. Chem.*, 250, 3814, 1975.
56. **Ohta, Y. and Ribbons, D. W.,** Bacterial metabolism of resorcinylic compounds, *Eur. J. Biochem.*, 61, 259, 1976.
57. **Neujahr, H. Y. and Gaal, A.,** *Eur. J. Biochem.*, 35, 386, 1973.
58. **Neujahr, H. Y. and Gaal, A.,** Phenol hydroxylase from yeast, *Eur. J. Biochem.*, 58, 351, 1975.
59. **Michalover, J. L. and Ribbons, D. W.,** *Biochem. Biophys. Res. Commun.*, 55, 1102, 1973.
60. **Groseclose, E. E. and Ribbons, D. W.,** 3-Hydroxybenzoate-6-hydroxylase from *Pseudomonas aeruginosa, Biochem. Biophys. Res. Commun.*, 55, 897, 1973.
61. **Premakumar, R., Subba Rao, P. V., Sreeleela, N. S., and Vaidyanathan, C. S.,** M-hydroxybenzoic acid-4-hydroxylase from *Aspergillus niger, Can. J. Biochem.*, 47, 825, 1969.
62. **Premakumar, R., Subba Rao, P. V., and Vaidyanathan, C. S.,** *Ind. J. Biochem. Biophys.*, 10, 184, 1973.
63. **Ashok Kumar, A., Vaidyanathan, C. S., and Appaji Rao, N.,** Mechanism of aromatic hydroxylation: the kinetic mechanism and the involvement of superoxide anions in the reaction catalysed by M-hydroxybenzoate-4-hydroxylase from *Aspergillus niger, Ind. J. Biochem. Biophys.*, 15, 5, 1978.
64. **Elsmore, A. E. and Hopper, D. J.,** The purification and properties of 4-hydroxyisopthalate hydroxylase from *Pseudomonas putida* NCIB 9866, *Eur. J. Biochem.*, 76, 197, 1977.
65. **Blakley, E. R.,** *Can. J. Microbiol.*, 18, 1247, 1972.
66. **Hareland, W. A., Crawford, R. L., Chapman, P. J., and Dagley, S.,** Metabolic function and properties of 4-hydroxyphenyl acetic acid-1-hydroxylase from *Pseudomonas acidovorans, J. Bacteriol.*, 121, 272, 1975.
67. **Gunsalus, I. C., Pederson, T. C., and Sligar, S. G.,** Oxygenase catalysed biological hydroxylation, *Ann. Rev. Biochem.*, 44, 377, 1975.
68. **Gunsalus, I. C., Meeks, J. R., Limpscomb, J. D., Debrunner, P., and Munck, E.,** *Molecular Mechanisms of Oxygen Activation*, Hayaishi, O., Ed., Academic Press, New York, 1974, 559.
69. **Ullrich, V. and Deppel, W.,** *The Enzymes*, Vol. 12, Boyer, P. D., Ed., Academic Press, New York, 1975, 253.
70. **Daly, J. W., Jerina, D. M., and Witkop, B.,** Arene oxide and the NIH shift; the metabolism, toxicity and carcinogenicity of aromatic compounds, *Experientia*, 28, 1129, 1972.
71. **Jerina, D. M. and Daly, J. W.,** Arene oxide: a new concept of drug metabolism, *Science*, 185, 573, 1974.
72. **Ullrich, V., Ruf, H. H., and Wende, P.,** The structure and mechanism of cytochrome P-450, *Croatica Chem. Acta*, 49, 213, 1977.
73. **Coon, M. J., White, R. E., Nordblum, G. D., Ballou, D. P., and Guengerich, F. P.,** Highly purified liver microsomal cytochrome P-450: properties and catalytic mechanism, *Croatica Chem. Acta*, 49, 163, 1977.
74. **Vermilion, J. L., Ballou, D. P., Massey, V., and Coon, M. J.,** Separate roles of FMN and FAD in catalysis of liver microsomal NADPH cytochrome P-450 reductase, *J. Biol. Chem.*, 256, 266, 1981.
75. **Blake, R. C. and Coon, M. J.,** On the mechanism of action of cytochrome P-450 spectral intermediates in the reaction of P-450 with peroxy compounds, *J. Biol. Chem.*, 255, 4100, 1980.
76. **Gustafsson, J. A. and Bergman, J.,** Iodine- and chlorine-containing oxidation agents as hydroxylating catalysts in cytochrome P-450 dependent fatty acid hydroxylation reaction in rat liver microsomes, *FEBS Lett.*, 70, 276, 1976.

77. **Hrycay, E. G., Gustaffson, J. A., Sundberg, M. L., and Ernster, L.,** The involvement of cytochrome P-450 in hepatic microsomal steroid hydroxylation reactions supported by sodium periodate, sodium chlorite and organic hydroperoxides, *Eur. J. Biochem.*, 61, 43, 1976.
78. **Potts, J. R. M., Weklych, R., and Conn, E. E.,** The 4-hydroxylation of cinnamic acid by sorghum microsomes and the requirement for cytochrome P-450, *J. Biol. Chem.*, 249, 5019, 1974.
79. **Diesperger, H. and Sandermann, H., Jr.,** Evidence for an electrophilic intermediate in microsomal hydroxylation of cinnamic acids in plants, *FEBS Lett.*, 85, 333, 1978.
80. **Murphy, G., Vogel, G., Kippahl, G., and Lynen, F.,** Patulin biosynthesis: the role of mixed function oxidases in the hydroxylation of m-cresol, *Eur. J. Biochem.*, 49, 443, 1974.
81. **Murphy, G. and Lynen, F.,** Patulin biosynthesis: the metabolism of m-hydroxybenzyl alcohol and m-hydroxybenzaldehyde by particulate prep. from *Penicillium patulum, Eur. J. Biochem.*, 58, 467, 1975.
82. **Sugumaran, M. and Vaidyanathan, C. S.,** Microsomal hydroxylation of phenyl acetic acid by *Aspergillus niger, FEMS Microbiol. Lett.*, 5, 427, 1979.
83. **Smith, R. V. and Rosazza, J. P.,** Microbial models of mammalian metabolism, aromatic hydroxylation, *Arch. Biochem. Biophys.* 161, 551, 1974.
84. **Smith, R. V., Humphrey, D. W., Engel, K. O., and Rosazza, J. P.,** Microbial models of mammalian metabolism: aromatic hydroxylation of isomeric xylenes, *Appl. Environ. Microbiol.*, 31, 448, 1976.
85. **Smith, R. V. and Rosazza, J. P.,** Microbial systems for the study of the biotransformations of drugs, *J. Pharm. Sci.*, 64, 1737, 1975.
86. **Goldstein, M., Joh, T. H., and Garvey, T. Q.,** Kinetic studies of the enzymatic dopamine-B-hydroxylation reaction, *Biochemistry*, 7, 2724, 1968.
87. **Friedman, S. and Kaufman, S.,** An electron paramagnetic study of 3,4-dihydroxyphenylethylamine-B-hydroxylase, *J. Biol. Chem.*, 241, 2256, 1966.
88. **Taylor, K. B.,** Dopamine-B-hydroxylase: stereochemical course of the reaction, *J. Biol. Chem.*, 249, 454, 1974.
89. **Lindblad, B., Lindstedt, G., and Rundgren, M.,** The mechanism of enzymic formation of homogentisate from *p*-hydroxyphenyl pyruvate, *J. Am. Chem. Soc.*, 92, 7446, 1970.
90. **Lindblad, B., Lindstedt, G., Lindstedt, S., and Rundgren, M.,** Purification and some properties of human 4-hydroxyphenyl pyruvate dioxygenase, *J. Biol. Chem.*, 252, 5073, 1977.
91. **Lindstedt, S., Odelhog, B., and Rundgren, M.,** Purification and some properties of 4-hydroxyphenyl pyrvuate dioxygenase from *Pseudomonas* sp., *Biochemistry*, 16, 3369, 1977.
92. **Nakai, C., Nozaki, M., Hayaishi, O., Saito, I., and Matusuura, T.,** Studies on a possible reaction intermediate of *p*-hydroxyphenyl pyruvate dioxygenase, *Biochem. Biophys. Res. Commun.*, 67, 590, 1975.
93. **Rundgren, M.,** Steady-state kinetics of 4-hydroxyphenyl pyruvate dioxygenase from human liver, *J. Biol. Chem.*, 252, 5094, 1977.
94. **Wada, G. H. and Fellman, J. H.,** 2,4,5-Trihydroxyphenylacetic acid: a metabolite of L-3,4-dihydroxyphenylalanine, *Biochemistry*, 12, 5212, 1973.
95. **Gibson, D. T.,** Microbial degradation of hydrocarbons, *Zbl. Bakt. Hyg. I. Abt. Orig.*, B162, 157, 1976.
96. **Gibson, D. T., Cardini, G. E., Maseles, F. C., and Kallio, R. E.,** Incorporation of oxygen-18 into benzene by *Pseudomonas putida, Biochemistry*, 9, 1631, 1970.
97. **Axcell, B. C. and Geary, P. J.,** Purification and some properties of a soluble benzene-oxidising system from a strain of *Pseudomonas, Biochem. J.*, 146, 173, 1975.
98. **Axcell, B. C. and Geary, P. J.,** The metabolism of benzene by bacteria: purification and some properties of the enzyme cis 1,2-dihydroxy-cyclohexa-3,5-diene (nicotinamide adenine nucleotide) oxidoreductase (cis-benzene glycol dehydrogenase), *Biochem. J.*, 136, 927, 1973.
99. **Yeh, W. K., Gibson, D. T., and Liu, T. N.,** Toluene dioxygenase: a multicomponent enzyme system, *Biochem. Biophys. Res. Commun.*, 78, 401, 1977.
100. **Rogers, J. E. and Gibson, D. T.,** Purification and properties of cis-toluene dihydrodiol dehydrogenase from *Pseudomonas putida, J. Bacteriol.*, 130, 1117, 1977.
101. **Subramanian, V., Sugumaran, M., and Vaidyanathan, C. S.,** Double hydroxylation reactions in microorganisms, *J. Indian Inst. Sci.*, 60, 143, 1978.
102. **Reiner, A. M. and Hegeman, G. D.,** Metabolism of benzoic acids by bacteria. Accumulation of (−)-3,5-cyclohexodiene-1,2-diol-1-carboxylic acid by a mutant strain of *Alcaligenes eutrophus, Biochemistry*, 10, 2530, 1971.
103. **Yamaguchi, M., Yamauchi, T., and Fujisawa, H.,** Studies on mechanisms of double hydroxylations: evidence for participation of NADH-cytochrome *c* reductase in the reaction of benzoate 1,2-dioxygenase (benzoate hydroxylase), *Biochem. Biophys. Res. Commun.*, 67, 264, 1975.
104. **Yamaguchi, M. and Fujisawa, H.,** Purification and characterization of an oxygenase component in benzoate 1,2-dioxygenase system from *Pseudomonas arvilla* c-1, *J. Biol. Chem.*, 255, 5058, 1980.
105. **Reiner, A. M.,** Metabolism of aromatic compounds in bacteria: purification and properties of the catechol family enzymes: 3,5-cyclohexadiene-1,2-diol-1-carboxylic acid (NAD^+) oxidoreductase (decarboxylating), *J. Biol. Chem.*, 247, 4960, 1972.

106. **Davey, J. F. and Gibson, D. T.,** Bacterial metabolism of para- and meta-xylene. Oxidation of a methyl substituent, *J. Bacteriol.*, 119, 923, 1974.
107. **DeFrank, J. J. and Ribbons, D. W.,** The *p*-cymene pathway in *Pseudomonas putida* PL: isolation of a dihydrodiol accumulation by a mutant, *Biochem. Biophys. Res. Commun.*, 70, 1129, 1976.
108. **DeFrank, J. J. and Ribbons, D. W.,** *p*-Cymene pathway in *Pseudomonas putida:* ring cleavage of 2,3-dihydroxy-*p*-cumate and subsequent reaction, *J. Bacteriol.*, 129, 1356, 1977.
109. **Keyser, P. and Ribbons, D. W.,** *Abstr. Annu. Meet. Am. Soc. Microbiol.*, 160, 1975.
110. **Bachofer, R., Lingens, F., and Schaffer, W.,** Conversion of aniline into pyrocatechol by a *Nocardia* Incorporation of oxygen, *FEBS Lett.*, 50, 288, 1975.
111. **Kobayashi, S., Kuno, S., Itada, N., Hayaishi, O., Kozuka, S., and Oae, S.,** ^{18}O studies on anthranilate hydroxylase. A novel mechanism of double hydroxylation, *Biochem. Biophys. Res. Commun.*, 16, 556, 1964.
112. **Kobayashi, S. and Hayaishi, O.,** Anthranilic acid conversion to catechol (*Pseudomonas*), in *Methods in Enzymology,* Vol. 17, Academic Press, New York, 1970, 505.
113. **Boyland, E.,** *Biochem. Soc. Symp.*, 5, 40, 1950.
114. **Boyland, E. and Sims, P.,** Metabolism of polycyclic compounds. The metabolism of 9,10-epoxy-9,10-dihydrophenanthrene in rats, *Biochem. J.*, 95, 788, 1965.
115. **Boyland, E. and Sims, P.,** The metabolism of benz(a) anthracene and dibenz(a,h)anthracene and their 5,6-epoxy-5,6-dihydroderivatives by rat liver, *Biochem. J.*, 97, 7, 1965.
116. **Jerina, D. M., Daly, J. W., Witkop, B., Zaltzman-Nirenberg, P., and Udenfriend, S.,** The role of arene oxide-oxepin system in the metabolism of aromatic substrates: formation of 1,2-naphthalene oxide from naphthalene by liver microsomes, *J. Am. Chem. Soc.*, 90, 6525, 1968.
117. **Jerina, D. M., Daly, J. W., Witkop, B., Zaltzman-Nirenberg, P., and Udenfriend, S.,** Role of arene oxide-oxepin system in the metabolism of aromatic substrates: *in vitro* conversion of benzene oxide to a premercapturic acid and a dihydrodiol, *Arch. Biochem. Biophys.*, 128, 176, 1968.
118. **Lu, A. Y. H., Ryan, D., Jerina, D. M., Daly, J. W., and Levin, W.,** Liver microsomal epoxide hydrase-solubilization, purification and characterisation, *J. Biol. Chem.*, 250, 8283, 1975.
119. **Lu, A. Y. H., Jerina, D. M., and Levin W.,** Liver microsomal epoxide hydrase: hydration of alkene and arene oxides by membrane bound and purified enzymes, *J. Biol. Chem.*, 252, 3715, 1977.
120. **Ayengar, P. K., Hayaishi, O., Nakajima, M., and Tomida, I.,** Enzymic aromatization of 3,5-cyclo-hexadiene-1,2-diol, *Biochim. Biophys. Acta,* 33, 111, 1959.
121. **Cerniglia, C. and Gibson, D. T.,** Metabolism of naphthalene by *Cunninghamella elegans, Appl. Environ. Microbiol.*, 34, 363, 1977.
122. **Carniglia, C. and Gibson, D. T.,** Metabolism of naphthalene by cell extracts of *Cunninghamella elegans, Arch. Biochem. Biophys.*, 186, 121, 1978.
123. **Krishnamurthy, H. G. and Simpson, F. J.,** Degradation of rutin by *Aspergillus flavus:* studies with oxygen-18 on the action of a dioxygenase on quercetin, *J. Biol. Chem.*, 245, 1467, 1970.
124. **Oka, T., Simpson, F. J., Child, J. J., and Mills, C.,** Degradation of rutin by *Aspergillus flavus:* purification of the dioxygenase-quercetinase, *Can. J. Microbiol.*, 17, 111, 1971.
125. **Oka, T., Simpson, F. J., and Krishnamurthy, H. G.,** Quercetinase, a dioxygenase containing copper, *Biochem. Biophys. Res. Commun.*, 43, 1, 1971.
126. **Oka, T., Simpson, F. J., and Krishnamurthy, H. G.,** Degradation of rutin by *Aspergillus flavus:* studies on specificity, inhibition and possible reaction mechanisms of quercetinase, *Can. J. Microbiol.*, 18, 493, 1972.
127. **Jeffrey, A. M., Knight, M., and Evans, W. C.,** The bacterial degradation of flavonoids. Hydroxylation of the A-ring of taxifalin by a soil *Pseudomonas, Biochem. J.*, 130, 373, 1972.
128. **Jeffrey, A. M., Jernia, D. M., Self, R., and Evans, W. C.,** Oxidative fission of the A-ring of dihydro-gossypetin by *Pseudomonas* sp., *Biochem. J.*, 130, 383, 1972.
129. **Scott, A. I. and Beadling, L. C.,** Biosynthesis of patulin. Dehydrogenase and dioxygenase enzymes of *Penicillium patulum, Bioorg. Chem.*, 3, 281, 1974.
130. **Lack, L.,** The enzymic oxidation of gentisic acid, *Biochim. Biophys. Acta,* 34, 117, 1959.
131. **Sugiyama, S., Yano, K., and Arima, K.,** *Bull. Agric. Chem. Soc. Jpn.*, 24, 249, 1966.
132. **Crawford, R. L., Hutton, S. W., and Chapman, P. J.,** Purification and properties of gentisate 1,2-dioxygenase from *Moraxella osloensis, J. Bacteriol.*, 121, 794, 1975.
133. **Adachi, K., Iwayama, Y., Tanioka, H., and Takeda, Y.,** Purification and properties of homogentisate oxygenase from *Pseudomonas fluorescens, Biochim. Biophys. Acta,* 118, 88, 1966.
134. **Chapman, P. J. and Dagley, S.,** Oxidation of homogentisic acid by cell-free extracts of a *Vibrio, J. Gen. Microbiol.*, 28, 251, 1962.
135. **Crawford, R. L.,** Degradation of homogentisate by strains of *bacillus* and *Moraxella, Can. J. Microbiol.*, 22, 276, 1976.
136. **Sugumaran, M. and Vaidyanathan, C. S.,** Affinity chromatography of homogentisate-1,2-dioxygenase from *Aspergillus niger, FEMS Microbiol. Lett.*, 4, 343, 1978.

137. **Durand, R. and Zenk, M. H.,** Enzymes of the homogentisate ring cleavage in cell suspension cultures of higher plants, *FEBS Lett.*, 39, 218, 1974.
138. **Suda, M. and Takeda, Y.,** Metabolism of tyrosine: homogentisicase, *J. Biochem.*, 37, 381, 1950.
139. **Knox, W. E. and Edwards, S. W.,** (i) Homogentisate oxidase of liver; (ii) properties of maleylacetoacetate, the initial product of homogentisate oxidation in liver, *J. Biol. Chem.*, 216, pp. 479 and 489, 1955.
140. **Tokuyama, K.,** Studies on Homogentisicase: (i) purification and the role of ferrous ion in the enzyme action; (ii) stoichiometry of the reactions and nature of reactive groups; (iii) kinetic studies on the enzyme action, *J. Biochem.*, 46, pp. 1379, 1453, and 1559, 1959.
141. **Takemori, S., Furuya, E., Mihara, K., and Katagiri, M.,** Bovine liver homogentisicase: apo and reconstituted holoenzymes, *Eur. J. Biochem.*, 6, 411, 1968.
142. **Koontz, W. A. and Shiman, R.,** Beef kidney 3-hydroxyanthranilic acid oxygenase: purification, characterization and analysis of the assay, *J. Biol. Chem.*, 251, 368, 1976.
143. **Fischer, N. and Dreiding, A. S.,** Biosynthesis of betalaines: on the cleavage of the aromatic ring during the enzymatic transformation of dopa into betalanic acid, *Helv. Chim. Acta*, 55, 649, 1972.
144. **Impellizzeri, G. and Piatelli, M.,** Biosynthesis of indicaxanthin in *Opuntiaficus-indica* fruits, *Phytochemistry*, 11, 2499, 1972.
145. **Saito, K. and Komamine, A.,** Biosynthesis of stizolobinic acid and stizolobic acid in higher plants. Stizolobinic acid synthase and stizolobinic acid synthase, new enzymes which catalyze the reaction sequences leading to the formation of stizolobinic acid and stizolobic acid from 3,4-dihydroxyphenylalanine in *Stizolobium hassjoo*, *Eur. J. Biochem.*, 82, 385, 1978.
146. **Von Ardenne, R., Dopp, H., Musso, H., and Steglich, W.,** *Z. Naturforsch.*, 29C, 637, 1974.
147. **Hurley, L. H., Gairola, C., and Das, N. V.,** Pyrrolo (1,4)-benzodiazepine antibiotics. Biosynthesis of the antitumor antibiotic 11-dimethyltomaymycin and its biologically inactive metabolite oxotomaymycin by *Streptomyces achromogenes*, *Biochemistry*, 15, 3760, 1976.
148. **Seidman, M. M., Toms, A., and Wood, J. M.,** Influence of sidechain substituents on the position of cleavage of the benzene ring by *Pseudomonas fluorescens*, *J. Bacteriol.*, 97, 1192, 1969.
149. **Dagley, S., Chapman, P. J., and Gibson, D. T.,** The metabolism of β-phenylpropionic acid by an *Achromobacter*, *Biochem. J.*, 97, 643, 1965.
150. **Adachi, K., Takeda, Y., Sench, S., and Kita, H.,** Metabolism of *p*-hydroxyphenylacetic acid in *Pseudomonas ovalis*, *Biochim. Biophys. Acta*, 93, 483, 1964.
151. **Jamaluddin, M. P.,** Purification and properties of homoprotocatechuate-2,3-dioxygenase from *Bacillus stearothermophilus*, *J. Bacteriol.*, 129, 690, 1977.
152. **Krishnan Kutty, R., Abitha Devi, N., Veeraswamy, M., Ramesh, S., and Subba Rao, P. V.,** Degradation of (±)-synephrine by *Arthrobacter synephrinum* — oxidation of 3,4-dihydroxyphenylacetate to 2-hydroxy, 5-carboxymethyl muconate semialdehyde, *Biochem. J.*, 167, 163, 1977.
153. **Dagley, S. and Wood, J. M.,** Oxidation of phenylacetic acid by a *Pseudomonas*, *Biochim. Biophys. Acta*, 99, 383, 1965.
154. **Kita, H.,** Crystallisation and some properties of 3,4-dihydroxyphenylacetate 2,3-oxygenase from *Pseudomonas ovalis*, *J. Biochem.*, 58, 116, 1965.
155. **Kita, H., Kamimoto, M., Senoh, S., Adachi, K., and Takeda, Y.,** Crystallisation and some properties of 3,4-dihydroxyphenylacetate-2,3-oxygenase, *Biochem. Biophys, Res. Commun.*, 18, 66, 1965.
156. **Kamimoto, M.,** Studies on 3,4-dihydroxyphenylacetate 2,3-dioxygenase. I. Role of iron substrate binding and some other properties, *J. Biochem.*, 74, 1049, 1973.
157. **Kamimoto, M. and Senoh, S.,** Studies on 3,4-dihydroxyphenylacetate-2,3-dioxygenase: role of sulfhydryl groups and subunit structure, *J. Biochem.*, 75, 321, 1974.
158. **Subba Rao, P. V., Friting, B., Vose, J. R., and Towers, G. H. N.,** An aromatic 3,4-oxygenase from *Telletiopsis washingtonensis*. Oxidation of 3,4-dihydroxyphenylacetic acid to *p*-carboxy methyl muconolactone, *Phytochemistry*, 10, 51, 1971.
159. **Que, L., Widom, J., and Crawford, R. L.,** 3,4-Dihydroxyphenylacetate 2,3-dioxygenase: a manganese(II) deoxygenase from *Bacillus brevis*, *J. Biol. Chem.*, 256, 10,941, 1981.
160. **Dagley, S. and Patel, M. D.,** Oxidation of *p*-cresol and related compounds by a *Pseudomonas*, *Biochem. J.*, 66, 227, 1957.
161. **Ono, K., Nozaki, M., and Hayaishi, O.,** Purification and some properties of protocatechuate 3,4-dioxygenase, *Biochim. Biophys. Acta*, 220, 224, 1970.
162. **Zabinski, R., Munck, M., Champion, P. M., and Wood, J. M.,** Kinetic and Mossbauer studies on the mechanism of protocatechuic acid 4,5-oxygenase, *Biochemistry*, 11, 3212, 1972.
163. **Crawford, R. L.,** Novel pathway for degradation of protocatechuic acid in *Bacillus* sp., *J. Bacteriol.*, 121, 531, 1975.
164. **Evans, W. C.,** Oxidation of phenol and benozic acid by some soil bacteria, *Biochem. J.*, 41, 373, 1947.
165. **Stainer, R. Y. and Ingraham, J. L.,** Protocatechuic acid oxidase, *J. Biol. Chem.*, 210, 799, 1954.
166. **Bull, C. and Ballou, D. P.,** Purification and properties of protocatechuate-3,4-dioxygenase from *Pseudomonas putida:* a new iron to subunit stoichiometry, *J. Biol. Chem.*, 256, 12,673, 1981.

167. **Fujisawa, H. and Hayaishi, O.,** Protocatechuic acid 3,4-dioxygenase: crystallization and characterization, *J. Biol. Chem.*, 243, 2673, 1968.
168. **Hou, C. T., Lillard, M. O., and Schwartz, R. D.,** Protocatechuate 3,4-dioxygenase from *Acinetobacter calcoaceticus, Biochemistry,* 15, 582, 1976.
169. **Cain, R. B. and Cartwright, N. J.,** On the properties of some aromatic ring-opening enzyme of species of the genus *Nocardia, Biochim. Biophys. Acta,* 37, 197, 1960.
170. **Cain, R. B., Bilton, R. F., and Darrah, J. A.,** The metabolism of aromatic acids by microorganisms: metabolic pathways in fungi, *Biochem. J.*, 108, 797, 1968.
171. **Satyashur, K. A., Rao, K. T., Lipscomb, J. D., and Wood, J. M.,** Preliminary crystallographic study of protocatechuate 3,4-dioxygenase from *Pseudomonas aeruginosa, J. Biol. Chem.,* 255, 10015, 1980.
172. **Kohlmiller, N. A. and Howard, J. B.,** The primary structure of α-subunit of protocatechuate 3,4-dioxygenase. I. Isolation and sequence of the tryptic peptides, *J. Biol. Chem.*, 254, 7302, 1979.
173. **Kohlmiller, N. A. and Howard, J. B.,** The primary structure of the α-subunit of protocatechuate 3,4-dioxygenase. II. Isolation and sequence of overlap peptides and complete sequence, *J. Biol. Chem.*, 254, 7309, 1979.
174. **Fujisawa, H., Uyeda, M., Kojima, Y., Nozaki, M., and Hayaishi, O.,** Protocatechuate 3,4-dioxygenase. II. Electron spin resonance and spectral studies on interaction of substrates and enzymes, *J. Biol. Chem.*, 247, 4414, 1972.
175. **Yoshida, R., Hori, K., Fujiwara, M., Saeki, Y., Kagamiyama, H., and Nozaki, M.,** Non-identical subunits of protocatechuate 3,4-dioxygenase, *Biochemistry,* 15, 4048, 1976.
176. **Ribbons, D. W.,** Bacterial oxidation of 2,3-dihydroxy benzoic acid — a new oxygenase, *Biochem. J.*, 99, 30P, 1966.
177. **Madhyastha, K. M., Rangachari, P. N., Raghavendra Rao, M., and Bhattacharya, P. K.,** Microbial transformations of terpenes. XV. Enzyme systems in the catabolism of *p*-cymene in PL-strain, *Ind. J. Biochem.*, 5, 167, 1968.
178. **DeFrank, J. J. and Ribbons, D. W.,** *p*-Cymene pathway in *Pseudomonas putida:* ring cleavage of 2,3-dihydroxy-*p*-cumate and subsequent reactions, *J. Bacteriol.*, 129, 1365, 1977.
179. **Sharma, H. K. and Vaidyanathan, C. S.,** A new mode of ring cleavage of 2,3-dihydroxybenzoic acid in *Tecoma stans* (L). Purification and properties of 2,3-dihydroxybenzoate 2,3-oxygenase, *Eur. J. Biochem.*, 56, 163, 1975.
180. **Hayaishi, O., Katagiri, M., and Rothberg, S.,** Mechanism of the pyrocatechase reaction, *J. Am. Chem. Soc.*, 77, 5450, 1955.
181. **Kojima, Y., Fujisawa, H., Nakagawa, A., Nakazawa, T., Kanetsuna, F., Taniuchi, H., Nozaki, M., and Hayaishi, O.,** Studies on pyrocatechase. I. Purification and spectral properties, *J. Biol. Chem.*, 242, 3270, 1967.
182. **Nakazawa, A., Kojima, Y., and Taniuchi, H.,** Purification and properties of pyrocatechase from *Pseudomonas fluorescens, Biochim. Biophys. Acta,* 147, 189, 1967.
183. **Patel, R. N., Hou, C. T., Felix, A., and Lillard, M. O.,** Catechol 1,2-dioxygenase from *Acinetobacter calcoaecticus:* purification and properties, *J. Bacteriol.*, 127, 536, 1976.
184. **Nakagawa, H., Inoue, H., and Takeda, Y.,** Characteristics of catechol oxygenase from *Brevibacterium fuseum, J. Biochem.*, 54, 65, 1963.
185. **Fujiwara, M., Golovleva, L. A., Saeki, Y., Nozaki, M., and Hayaishi, O.,** Extradiol cleavage of 3-substituted catechols by an intradiol dioxygenase, pyrocatechase from a Pseudomonas, *J. Biol. Chem.*, 250, 4848, 1975.
186. **Que, L., Jr.,** Extradiol cleavage of *o*-aminophenol by pyrocatechase, *Biochem. Biophys. Res. Commun.*, 84, 123, 1978.
187. **Nozaki, M., Kagamiyama, H., and Hayaishi, O.,** Metapyrocatechase. I. Purification, crystallization and some properties, *Biochem. Z.*, 338, 528, 1963.
188. **Taniuchi, H., Kojima, Y., Kanetsuna, F., Ochiai, H., and Hayaishi, O.,** Mechanism of autooxidation of oxygenases, *Biochem. Biophys. Res. Commun.*, 8, 97, 1962.
189. **Catelani, D., Fiecchi, A., and Galli, E.,** (+)-γ-Carboxymethyl-γ-methyl-Δ^{α}-butenolide, the 1,2-ring fission product of 4-methylcatechol by *Pseudomonas desmolyticum, Biochem., J.*, 121, 89, 1971.
190. **Hashimoto, K.,** Oxidation of phenols by yeast. II. Oxidation of cresols by *Candida tropicalis, J. Gen. Appl. Microbiol.*, 19, 171, 1973.
191. **Hashimoto, M.,** Oxidation of phenols by yeast. I. A new oxidation product from *p*-cresol by an isolated strain of yeast, *J. Gen. Appl. Microbiol.*, 16, 1, 1970.
192. **Sugumaran, M. and Vaidyanathan, C. S.,** Metabolism of aromatic compounds, *J. Ind. Inst. Sci.*, 60, 57, 1978.
193. **Tack, B. F., Chapman, P. J., and Dagley, S.,** Metabolism of gallic acid and syringic acid by *Pseudomonas putida, J. Biol. Chem.*, 247, 6438, 1972.

194. **Chapman, P. J. and Ribbons, D. W.,** Metabolism of resorcinylic compounds by bacteria: orcinol pathway in *Pseudomonas putida* and alternative pathways for resorcinol catabolism in *P. putida, J. Bacteriol.*, 125, pp. 975 and 985, 1976.
195. **Munnecke, D. M. and Hsieh, D. P. H.,** Pathway for microbial metabolism of parathion, *Appl. Microbiol.*, 31, 63, 1976.
196. **Chamberlain, E. M. and Dagley, S.,** The metabolism of thymol by a *Pseudomonas, Biochem. J.*, 110, 755, 1968.
197. **Sparnins, V. L., Anderson, J. J., Omans, J., and Dagley, S.,** Degradation of 4-hydroxyphenylacetic acid by *Trichosporon cutaneum, J. Bacteriol.*, 136, 449, 1978.
198. **Dagley, S.,** in *Soil Biochemistry*, McLaren, A. D. and Peterson, G. H., Eds., Marcel Dekker, New York, 1967, 287.
199. **Brown, D. G.,** *The Chemistry and Biochemistry of Amino Acids, Peptides and Proteins*, Vol. 3, Weinstein, B., Ed., Marcel Dekker, New York, 1974, 245.
200. **Gibson, D. T.,** Microbial degradation of aromatic compounds, *Science*, 161, 1093, 1968.
201. **Hayaishi, O. and Nozaki, M.,** Nature and mechanisms of oxygenases, *Science*, 164, 389, 1969.
202. **Towers, G. H. N. and Rao, P. V. S.,** Degradative metabolism of phenylalanine, tyrosine and dopa, *Recent Adv. Phytochem.*, 4, 1, 1972.
203. **Hayaishi, O., Nozaki, M., and Abbott, M. T.,** *Oxygenases: Dioxygenases in the Enzymes*, Boyer, P. D., Ed., Vol. 12, Academic Press, New York, 1975, 119.
204. **Sugumaran, M., Kishore, G., Subramanian, V., Mohan, V. P., and Vaidyanathan, C. S.,** Dioxygenases in higher plants — a review, *J. Ind. Inst. Sci.*, 59, 1, 1977.
205. **Wood, J. M., Crawford, R. L., Muenck, E., Zimmermann, R., Lipscomb, J. D., Stephens, R. S., Bromley, J. W., Que L., Jr., Howard, J. B., and Orme-Johnson, W. H.,** *J. Agric. Food Chem.*, 25, 698, 1977.
206. **Dagley, S.,** Microbial catabolism, the carbon cycle and environmental pollution, *Naturwissenschaften*, 65, 85, 1978.
207. **Hamilton, G. A.,** in *Molecular Mechanisms of Oxygen Activation*, Hayaishi, O., Ed., Academic Press, New York, 1974, 405.
208. **Kearns, D. R.,** Physical and chemical properties of singlet molecular oxygen, *Chem. Rev.*, 71, 395, 1971.
209. **Moro-Oka, Y. and Foote, C. S.,** Chemistry of superoxide ion. I. Oxidation of 3,5-di-*tert*-butylcatechol with KO_2, *J. Am. Chem. Soc.*, 98, 1510, 1976.
210. **Lee-Ruff, R., Lever, A. B. P., and Rigandy, J.,** The reaction of catechol and derivatives with potassium superoxide, *Can. J. Microbiol.*, 54, 1837, 1976.
211. **Nishinaga, A., Itahara, T., and Matsuura, T.,** *Bull. Chem. Soc. Jpn.*, 49, 3353, 1976.
212. **Hirata, F. and Hayaishi, O.,** Possible participation of superoxide anion in the intestinal tryptophan 2,3-dioxygenase reaction, *J. Biol. Chem.*, 246, 7825, 1971.
213. **Hirata, F. and Hayaishi, O.,** Indoleamine 2,3-dioxygenase. I. Superoxide as substrate, *J. Biol. Chem.*, 250, 5960, 1975.
214. **Ohnishi, T., Hirata, F., and Hayaishi, O.,** Indoleamine 2,3-dioxygenase: potassium superoxide as substrate, *J. Biol. Chem.*, 252, 4643, 1977.
215. **Hirata, F., Ohnishi, T., and Hayaishi, O.,** Indoleamine 2,3-dioxygenase; characterization and properties of enzyme-O_2^- complex, *J. Biol. Chem.*, 252, 4637, 1977.
216. **Hayaishi, O., Hirata, F., Ohnishi, T., Henry, J. P., Rosenthal, I., and Kotch, A.,** Indoleamine 2,3-dioxygenase — incorporation of $^{18}O_2$ into the reaction products, *J. Biol. Chem.*, 252, 3548, 1977.
217. **Boeseken, J. and Engelberts, R.,** *Proc. Acad. Sci. (Amsterdam)*, 34, 1292, 1931.
218. **Pandell, A. J.,** Enzyme-like aromatic oxidations. Metal-catalysed peracetic acid oxidation of phenol and catechol to *cis, cis*-muconic acid, *J. Org. Chem.*, 41, 3992, 1976.
219. **Grinstead, R. R.,** Metal-catalysed oxidation of 3,5-di-*t*-butylpyrocatechol and its significance in the mechanism of pyrocatechase action, *Biochemistry*, 3, 1308, 1964.
220. **Baldwin, J.,** Unpublished results, *J. Agric. Food Chem.*, 25, 698, 1977.
221. **Que, L., Jr., Lipscomb, J. D., Muenck, E., and Wood, J. M.,** Protocatechuate 3,4-dioxygenase. Inhibitor studies and mechanistic implications, *Biochim. Biophys. Acta*, 485, 60, 1977.
222. **Hamilton, G. A.,** Mechanisms of 2- and 4-electron oxidations catalysed by some metalloenzymes, *Adv. Enzymol.*, 32, 55, 1969.
223. **Fewson, C. A.,** in *Microbial Degradation of Xenobiotics and Recalcitrant Compounds*, Leisinger, T., Cook, A. M., Hutter, R., and Nuesch, J., Eds., Academic Press, New York, 1981, 141.
224. **Williams, P. A.,** in *Microbial Degradation of Xenobiotics and Recalcitrant Compounds*, Leisinger, T., Cook, A. M., Hutter, R., and Nuesch, J., Eds., Academic Press, New York, 1981, 97.

Chapter 6

REGULATION OF SECONDARY METABOLISM BY TRACE METALS

Eugene D. Weinberg

TABLE OF CONTENTS

I. INTRODUCTION

During the past 4 decades, it has become apparent that the synthesis of secondary metabolites as well as of structures required for cellular differentiation has a much narrower tolerance for range of environmental concentrations of specific trace metals and of P_i than does vegetative growth of the producer cells. Of the ten biologically important transition series metals (i.e., atomic numbers 23 to 30, 42, and 48), the quantity of iron generally is critical for secondary metabolic processes in prokaryotes and that of zinc in fungi.[1-3] Manganese concentration is consequential in secondary metabolism and differentiation of species of *Bacillus* and, in a few fungal systems, can be more important than that of zinc.[1-3] In both pro- and eukaryotic cells, vegetative growth is permitted throughout the range of 0.3 to 300 m*M* P_i; secondary metabolism and differentiation often are restricted by >10 m*M*.[2] Selected examples that illustrate these generalizations will be summarized in this chapter. A few studies have been concerned either with preferential acquisition of the key metals in post-exponential growth phase or with mechanisms whereby they might control secondary metabolism. These will be reviewed briefly.

II. PRESENT STATE OF KNOWLEDGE

Typical examples of suppression or stimulation of biosynthesis of secondary metabolites in prokaryotic systems by quantities of iron that, in many cases, have no effect on vegetative growth are contained in Figure 1.[4-22] In 8 of the 21 examples shown, however, the greatest yield is produced in cultures that contain insufficient iron for maximal cell multiplication. Note that in each of the 21 systems the maximal yield of secondary product occurs through a relatively small range of iron concentrations. Indeed, a pioneer worker in this field commented that "so narrow is the zone [of iron quantities] in which [diphtheria] toxin is obtained and so sharp the peak of maximal production that this single uncontrolled factor must have played a greater role in any previous experiments than specific conditions supposedly under investigation."[23]

Iron modulates production by prokaryotes of other extracellular enzymes and toxins in addition to the secondary metabolites illustrated in Figure 1. For example, in complex media, the optimal iron concentration for maximal formation of the lecithinase of *Clostridium perfringens,*[24] the enterotoxins of *Staphylococcus aureus,*[25] and of the neutrotoxins of *C. tetani,*[26] *Escherichia coli,*[27] and *Salmonella typhimurium*[27] was reported to be, respectively, 20, 100, 200, 600, and 600 μ*M*. In synthetic media, the maximal formation of the neutrotoxin of *C. tetani*[28] and of *C. botulinum*[29] occurred, respectively, at 1.0 and 5.4 μ*M* iron. Synthesis of an extracellular alkaline protease by *Vibrio alginolyticus* was completely suppressed by a powerful iron-binding agent, 1,10-phenanthroline.[30] In a study of 118 isolates of *Pseudomonas fluorescens,* formation of antifungal secondary metabolite(s) was suppressed by 62 μ*M* iron in 11%, unaffected in 16%, and enhanced in 73%.[31]

For efficient secondary metabolism and differentiation of actinomycetes, the quantity not only of iron but also of zinc is often critical. However, in nearly all fungal systems studied, although zinc concentration is very important in the control of the secondary processes, iron rarely is so. Examples of fungal and actinomycetal systems in which a much finer adjustment of zinc concentration is needed to obtain secondary metabolism and/or differentiation than to achieve growth of the producer cells are contained in Table 1.[8,9,32-53] Excellent vegetative growth of fungi generally occurs throughout a range of zinc quantities of two to three orders of magnitude, whereas the range of concentrations of the metal that permits high yield of a secondary product or of differentiated cells is usually less than one order of magnitude.

As mentioned above, the concentration of manganese consistently is critical in secondary metabolism and differentiation of strains of *Bacillus.* Although excellent vegetative growth

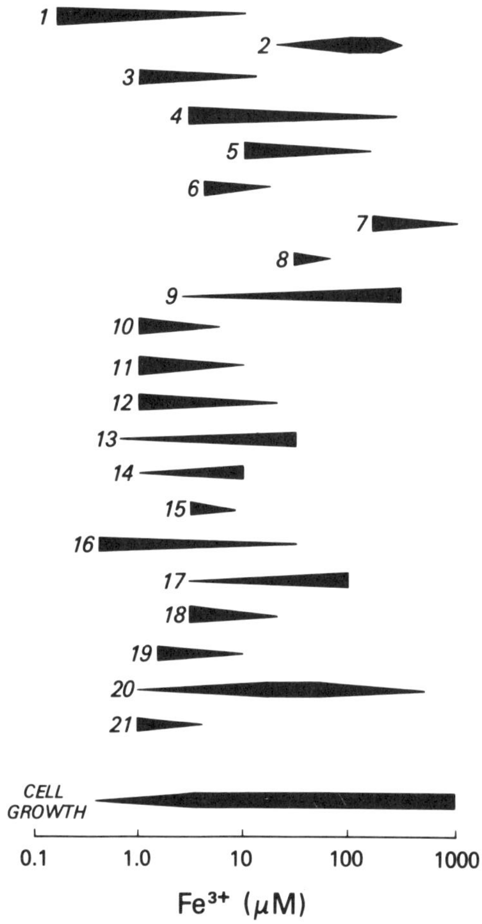

FIGURE 1. Modulation of yield of bacterial secondary metabolites by iron. Bases of arrows indicate amounts of iron that permit maximal yield of metabolites; arrow points denote quantities of iron at which minimal yield is obtained. Bar labeled "cell growth" specifies range of iron concentrations throughout which maximal growth of producer cells occurs; tapered portion of this bar indicates diminution of cell growth. The 21 arrows are arranged in alphabetical order with the seven Gram-positive species first followed by the Gram-negative species. Key: (1) *Corynebacterium diphtheriae* exotoxin,[4] (2) *Corynebacterium hydrocarboclastus* corynecin,[5] (3) *Listeria monocytogenes* hemolysin,[6] (4) *Streptomyces antibioticus* antimycin,[7] (5) *Streptomyces fradiae* neomycin,[8] (6) *Streptomyces griseus* streptomycin,[9] (7) *Streptomyces parvulus* actinomycin D,[10] (8) *Aeromonas hydrophila* exotoxin,[11] (9) *Chromobacterium violaceum* hydrogen cyanide,[12] (10) *Pseudomonas aeruginosa* alkaline protease,[13] (11) *Pseudomonas aeruginosa* elastase,[13] (12) *Pseudomonas aeruginosa* exotoxin A,[13] (13) *Pseudomonas aeruginosa* hydrogen cyanide,[14] (14) *Pseudomonas aeruginosa* pyocyanine,[15] (15) *Pseudomonas aeruginosa* pyoverdine,[16] (16) *Pseudomonas fluorescens* fluorescin,[17] (17) *Pseudomonas reptilovora* antibiotic,[18] (18) *Serratia marcescens* prodigiosin,[19] (19) *Shigella dysenteriae* exotoxin,[20] (20) *Vibrio cholerae* exotoxin,[21] (21) *Vibrio harveyi* luciferase.[22]

is attained at <0.5 μ*M*, to obtain good yields of either tripartite anthrax toxin,[54] antibiotics such as bacitracin,[55] mycobacillin,[56] subtilin,[57] or surfactin,[58] or of endospores,[59] at least 5.0 μ*M* manganese is required. Indeed, in species of *Bacillus,* even the processes of transfection, transformation, and viral synthesis are modulated by the concentration of this key metal.[1,3] Iron generally has a much less direct role; usually, it is required simply for maximal cell yield of the strain of *Bacillus* (see Reference 58).

Occasionally, manganese is important in a secondary metabolic system in prokaryotes other than *Bacillus* sp. For example, 1000 (but not 100) μ*M* manganese was required in synthetic medium for synthesis of a cholera toxin-like exotoxin by *Salmonella enteritidis.*[60]

Table 1
SECONDARY METABOLITES OR DIFFERENTIATED CELL FORMS WHOSE SYNTHESIS REQUIRES A FINER ADJUSTMENT OF ZINC CONCENTRATION THAN IS NECESSARY FOR MAXIMAL VEGETATIVE GROWTH OF THE PRODUCER MICROORGANISMS

Fungal products or structures		Actinomycete products
Aflatoxin[32]	Ochratoxin[41]	Actinomycin D[50]
Cynodontin[33]	Oogonia (Pythium)[42]	Candicidin[51]
Ergotamine[34]	Patulin[43]	Chloramphenicol[52]
Fusaric acid[35]	Penicillin[44]	Kanamycin[53]
Gentisyl alcohol[36]	Rubratoxin[45]	Neomycin[8]
Griseofulvin[37]	Sclerotia (Whetzelinia)[46]	Streptomycin[9]
Lysergic acid[38]	Vesicles (Puccinia)[47]	
Malformin[39]	Yeast→Mycelia (Candida)[48]	
Naphthazarin[40]	Zearalenone[49]	

This exceptionally high requirement for the metal might be due to a contaminant (e.g., chromium or vanadium) in the particular batch of manganous chloride employed in the study. Alternatively, the particular strain used might have had a genetic defect in manganese transport.

In some fungal systems, manganese is as, or more, critical than zinc in modulation of secondary metabolism and/or differentiation. For instance, the amount of each of the two metals is important in the control of aflatoxigenesis.[61] However, manganese is more consequential than is zinc in formation of malformin by *Aspergillus niger*[62] and in coremia synthesis by *Penicillium* sp.[63] In *A. niger,* 0.5 μM manganese suppresses citric acid production.[64]

The restricted quantitative requirement for key metal(s) for successful secondary metabolism and differentiation should be reflected in well-regulated control of uptake of that metal(s) during the post-exponential stage of the batch growth cycle. Evidence continues to accumulate in support of such control. For example, in *Candida utilis,* cells have a high capacity for zinc uptake only during lag phase and in late exponential phase (Figure 2).[65] During early and mid-exponential phase, the V_{max} of the transport system was 0.22 nmol Zn^{2+} per minute per milligram dry weight; in late exponential phase, it was increased 17-fold to 3.65 nmol Zn^{2+} per minute per milligram dry weight. Predictably, iron, whose concentration is not critical for fungal secondary metabolism, should be accumulated throughout the batch growth cycle at a steady state; this is, in fact, the case (Figure 2).[65]

In a prokaryote system, on the other hand, iron could be expected to be assimilated early in the growth cycle and again at the time that vegetative growth is slowing. Evidence for this prediction is contained in a study on iron uptake in the cyanobacterium *Agmenellum quadruplicatum* (Figure 3).[66] Similarly, in *Bacillus subtilis,* we would predict that manganese would be acquired in a cyclical manner; such is indeed the case. The metal is accumulated during exponential growth and then a period of equilibrium occurs during which the quantity per cell remains stable. At 3 hr into the sporulation sequence, manganese again is accumulated and eventually transferred into the developing spore.[67]

Inasmuch as ''key'' metal(s) are critical in secondary metabolism and differentiation in large taxonomic groups of microorganisms, it is tempting to seek a unitary mechanism of action of the metals. Because the known metabolic functions of manganese, iron, and zinc are so different from one another,[67-69] however, it has been difficult to formulate a unitary hypothesis. Nevertheless, in the case of iron the nutritional message is generally considered to be directed at the level of transcription. For example, in *Cornyebacterium diphtheriae*

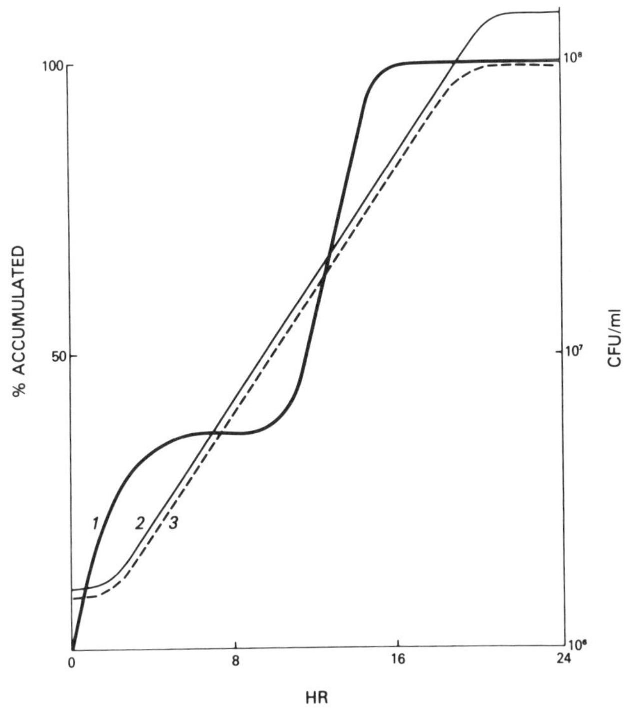

FIGURE 2. Accumulation of zinc and iron by cells of *Candida utilis* in batch culture. Curve 1 = zinc; curve 2 = cell growth; curve 3 = iron. Redrawn from data in Figure 1 of Reference 65.

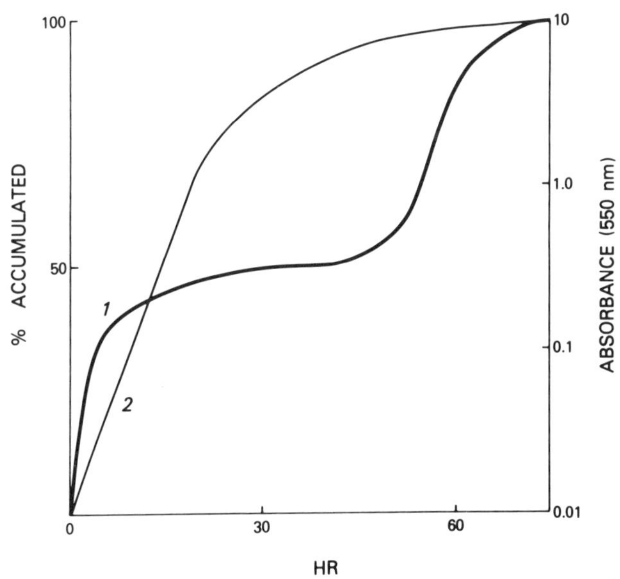

FIGURE 3. Accumulation of iron by cells of *Agmenellum quadruplicatum* in batch culture. Curve 1 = iron; curve 2 = cell growth. Redrawn from data in Figure 2 of Reference 66.

toxinogeny, the bacteria are believed to produce an iron-binding *tox* aporepressor.[70] In the presence of the metal, a repressor-iron complex would form and have affinity for the corynephage β*tox* operator locus. Under conditions of iron limitation, the repressor iron complex would dissociate and the diphtheria *tox* gene would become derepressed.[70] In systems in which iron is required for production of the specific secondary metabolic product, the metal might activate a derepressor rather than a repressor.

In the secondary systems in which the key metal stimulates the process at one concentration and inhibits at another, the mediator may consist of an allosteric repressor or derepressor protein that contains two metal binding sites. A model for such a proposal is that of the rat liver protein, fructose-1,6-bisphosphatase, whose activity is inhibited by 0.3 μ*M* and activated by 10 μ*M* zinc. At the low concentration, the metal binds solely to a high affinity site; at the higher amount, a second binding site is filled as well.[71]

It should be emphasized that the "correct" quantity of key metal essential for secondary metabolism must be present at a time in the history of the batch culture in which the cells are competent to utilize the element. If addition of the metal is delayed for several hours beyond the time at which it is needed, secondary metabolism and differentiation will be less efficient or may fail to occur at all.

Generally, the key metals are believed to participate in initiation but not termination of transcription of secondary metabolic genes. Nevertheless, termination is responsive to environmental messages. For example, in the *Pseudomonas aeruginosa* cyanogenic system, regulation appears to be dependent on cultural oxygen levels. The normal decline in oxygen during late exponential growth phase appears to be a signal (in addition to the appropriate amount of iron) to the cells to begin synthesizing the cyanogenic enzyme. The natural re-establishment of aerobic conditions early in stationary phase apparently results in repression of continued formation of this enzyme.[72]

As mentioned above, the range of concentrations of P_i that permit secondary metabolism and differentiation is far narrower in all taxonomic groups of microorganisms than is the range that permits excellent vegetative growth.[2,3] In one series of 26 systems compiled randomly,[2] the highest quantity of P_i that permitted maximal yield of metabolite or secondary process ranged from 0.01 to 10 m*M* with a mean value of 1.0 m*M* P_i. In contrast, excellent primary metabolism (i.e., vegetative growth) occurs throughout the range of 0.3 to 300 m*M* P_i.

High amounts of P_i prevent the formation of synthetases of secondary metabolism (see Reference 73) but, unlike iron, probably do not function directly to repress transcription. Numerous phosphatases are required in secondary metabolism, and the synthesis of alkaline phosphatases as well as some acid phosphatases is derepressed in low P_i environments. Moreover, nucleotides such as cyclic AMP, ATP, and 5′-AMP are even more potent than is 5.0 m*M* P_i in inhibiting candicidin synthesis; these materials might act by maintaining a high energy charge in the cells.[74]

III. FUTURE RESEARCH AND POSSIBLE APPLICATIONS

Observations on the exquisite sensitivity of secondary metabolism and differentiation to seemingly innocuous fluctuations in nutritional and environmental conditions such as those cited herein raise numerous questions for additional research. For example, what evolutionary factors caused the establishment of such a narrow tolerance of secondary processes to trace elements, and why does the "key" metal differ among *Bacillus* species, all other bacteria, and fungi? Furthermore, why does trace metal control function only at a precise stage in the life of the batch culture?

To what extent is secondary metabolism and differentiation of cells of such other taxonomic groups as green plants and vertebrate animals controlled by trace metals? Inasmuch as secondary processes finally can be studied in cell (rather than in tissue or organ) cultures of these groups, such problems can now be examined by cell biologists with the precision formerly available only to microbiologists.[75]

By manipulation of trace metals, P_i, and such other environmental agents as aeration and temperature that restrict secondary processes, industrial microbiologists obtain enhanced yields of desirable and lesser formation of unwanted secondary metabolites. In contrast, environmental and medical microbiologists rarely attempt to employ the restrictive agents in foods, waters, soils, or hosts, either to suppress yields of undesirable secondary products or to enhance formation of such useful metabolites as pesticides and herbicides. Nevertheless, there exists vigorous competition in foods, waters, soils, and hosts for such metals as iron.[76]

Knowledge of competition for iron recently has been applied to suppress bacterial and fungal pathogens of domestic plants.[77] Root-colonizing fluorescent pseudomonads can markedly increase yields of potato, sugar beet, and other useful crops by secreting iron-binding substances such as pseudobactin in iron-restricted soils. Fortunately, this molecule withholds iron from pathogenic microorganisms but not from the plants; it is now being used to convert "conducive" soils to "suppressive" soils.[78] Clearly, the controlled use of metal-binding agents in the prevention or management of infectious diseases by suppressing microbial growth or toxin formation is just beginning. Quite probably, continued research in this area will yield additional nutritionally and/or medicinally useful products for agriculture and medicine.

REFERENCES

1. **Weinberg, E. D.,** Biosynthesis of secondary metabolites: roles of trace metals, *Adv. Microb. Physiol.*, 4, 1, 1970.
2. **Weinberg, E. D.,** Mineral element control of microbial secondary metabolism, in *Microorganisms and Minerals*, Weinberg, E. D., Ed., Marcell Dekker, New York, 1977, 289.
3. **Weinberg, E. D.,** Biosynthesis of microbial metabolites: regulation by mineral elements and temperature, in *Overproduction of Microbial Products*, Krumphanzl, V., Sikyta, B., and Vanek, Z., Eds., Academic Press, London, 1982, 181.
4. **Kato, I.,** Relationship of bacterial structure to toxin production, in *Microbial Toxins I*, Ajl, S., Kadis, S., and Montie, T. C., Eds., Academic Press, New York, 1970, 401.
5. **Tonita, F., Nakano, H., and Suzucki, T.,** Further studies on production of chloramphenicol analogs (corynecins) from n-alkanes, *Agric. Biol. Chem.*, 38, 1673, 1974.
6. **Cowart, R. E. and Foster, B. G.,** The role of iron in the production of hemolysin by *Listeria monocytogenes*, *Curr. Microbiol.*, 6, 287, 1981.
7. **Neft, N. and Farley, T. M.,** Conditions influencing antimycin production by a *Streptomyces* species grown in chemically defined medium, *Antimicrob. Agents Chemother.*, 1, 274, 1972.
8. **Majumdar, M. K. and Majumdar, S. K.,** Effects of minerals on neomycin production by *Streptomyces fradiae*, *Appl. Microbiol.*, 13, 190, 1965.
9. **Chesters, C. G. C. and Rolinson, G. N.,** Effect of metals on streptomycin production, *J. Gen. Microbiol.*, 5, 559, 1951.
10. **Williams, W. K. and Katz, E.,** Development of a chemically defined medium for the synthesis of actinomycin D by *Streptomyces parvulus*, *Antimicrob. Agents Chemother.*, 11, 281, 1977.
11. **Riddle, L. M., Graham, T. E., and Amborski, R. L.,** Medium for the accumulation of extracellular hemolysin and protease by *Aeromonas hydrophila*, *Infect. Immun.*, 33, 728, 1981.
12. **Rodgers, P. B. and Knowles, C. J.,** Cyanide production and degradation during growth of *Chromobacterium violaceum*, *J. Gen. Microbiol.*, 108, 261, 1978.
13. **Sokol, P. A., Cox, D. C., and Iglewski, B. H.,** *Pseudomonas aeruginosa* mutants altered in their sensitivity to the effect of iron on toxin A or elastase yields, *J. Bacteriol.*, 151, 783, 1982.
14. **Castric, P. A.,** Hydrogen cyanide, a secondary metabolite of *Pseudomonas aeruginosa*, *Can. J. Microbiol.*, 21, 613, 1975.
15. **Kurachi, M.,** Studies on the biosynthesis of pyocyanine, *Bull. Inst. Chem. Res. Kiyoto Univ.*, 36, 188, 1958.

16. **Meyer, J. M. and Abdallah, M. A.,** The fluorescent pigment of *Pseudomonas fluorescens:* biosynthesis, purification and physiochemical properties, *J. Gen. Microbiol.*, 107, 319, 1978.
17. **Totter, J. R. and Mosely, F. T.,** Influence of concentration of iron on production of fluorescin by *Pseudomonas, J. Bacteriol.*, 65, 45, 1953.
18. **Martinez-Molina, E., Del Rio, L. A., and Olivares, J.,** Copper and iron as determinant factors of antibiotic production by *Pseudomonas reptilovora, J. Appl. Bacteriol.*, 41, 69, 1976.
19. **Waring, W. S. and Werkman, C. H.,** Iron requirements of heterotrophic bacteria, *Arch. Biochem.*, 1, 425, 1943.
20. **van Heyningen, W. E.,** The exotoxin of *Shigella dysenteriae,* in *Microbial Toxins II,* Kadis, S., Montie, T. C., and Ajl, S., Eds., Academic Press, New York, 1971, 255.
21. **Yancey, R. J. and Finkelstein, R. A.,** The effect of iron on toxigenesis of *Vibrio cholerae, Abstr. Annu. Meet. Am. Soc. Microbiol.*, B40, 1979.
22. **Makemson, J. C. and Hastings, J. W.,** Iron represses bioluminescence and affects catabolite repression of luminescence in *Vibrio harveyi, Curr. Microbiol.*, 7, 181, 1982.
23. **Mueller, J. H.,** The influence of iron on the production of diphtheria toxin, *J. Immunol.*, 42, 343, 1941.
24. **Ispolatovskaya, M. V.,** Type A *Clostridium perfringens* toxin, in *Microbial Toxins IIA,* Kadis, S., Montie, T. C., and Ajl, S., Eds., Academic Press, New York, 1971, 109.
25. **Morita, T. N., Patterson, J. E., and Woodburn, M. J.,** Magnesium and iron addition to casein hydrolysate medium for production of staphylococcal enterotoxins A, B, and C, *Appl. Environ. Microbiol.*, 38, 39, 1979.
26. **Latham, W. C., Bent, D. F., and Levine, L.,** Tetanus toxin production in the absence of protein, *Appl. Environ. Microbiol.*, 10, 146, 1962.
27. **Mesrobeneau, L. and Mesrobeneau, I.,** *Salmonella typhimurium* and *Escherichia coli* neurotoxins, in *Microbial Toxins IIA,* Kadis, S., Montie, T. C., and Ajl, S., Eds., Academic Press, New York, 1971, 301.
28. **van Heyningen, W. E. and Mellanby, J.,** Tetanus toxin, in *Microbial Toxins IIA,* Kadis, S., Montie, T. C., and Ajl, S., Eds., Academic Press, New York, 1971, 69.
29. **Siegel, L.,** Effect of iron on growth and toxin production by *Clostridium botulinum* Type A, *Curr. Microbiol.*, 6, 127, 1981.
30. **Hare, P., Long, S., Robb, F. T., and Woods, D. R.,** Regulation of exoprotease production by temperature and oxygen in *Vibrio alginolyticus, Arch. Microbiol.*, 130, 276, 1981.
31. **Hemming, B. C., Orser, C., Jacobs, D. L., Sands, D. C., and Strobel, G. A.,** The effects of iron on microbial antagonism by fluorescent pseudomonads, *J. Plant Nutr.*, 5, 683, 1982.
32. **Marsh, P. B., Simpson, M. E., and Trucksees, M. W.,** Effects of trace metals on the production of aflatoxins by *Aspergillus parasiticus, Appl. Microbiol.*, 30, 52, 1975.
33. **White, J. P. and Johnson, G. T.,** Zinc effects on growth and cynodontin production of *Helminthosporium cynodontis, Mycologia,* 63, 548, 1971.
34. **Stoll, A., Brack, A., Hofmann, A., and Kobel, H.,** Process for the Preparation of Ergotamine, U.S. Patent 2,809,920, 1957.
35. **Kalyanasundaraman, R. and Saraswathi-Devi, L.,** Zinc in the metabolism of *Fusarium vasinfectum* Atk., *Nature (London),* 1975, 945, 1955.
36. **Ehrensvaard, G.,** Some observations on aromatic biosynthesis in *Penicillium urticae* Bainer, *Exp. Cell Res.*, Suppl. 3, 102, 1955.
37. **Grove, J. F.,** Griseofulvin, in *Antibiotics II, Biosynthesis,* Gottlieb, D. and Shaw, P. D., Eds., Springer-Verlag, Berlin, 1967, 123.
38. **Rosazza, J. P., Kelleher, W. J., and Schwarting, A. E.,** Production of lysergic acid derivatives in submerged culture. IV. Inorganic nutrition studies with *Claviceps paspali, Appl. Microbiol.*, 15, 1270, 1967.
39. **Steenbergen, S. T. and Weinberg, E. D.,** Trace metal requirements for malformin biosynthesis, *Growth,* 32, 125, 1968.
40. **Kern, H., Naef-Roth, S., and Ruffner, F.,** Der Einfluss der Ernahrung auf die Bildung von Naphthazarinderivaten durch *Fusarium Martii* var. *pisi, Phytopathol. Z.*, 74, 272, 1972.
41. **Mingtan, L., Semeniuk, G., and Hesseltine, C. W.,** Conditions for production of ochratoxin A by *Aspergillus* species in a synthetic medium, *Appl. Microbiol.*, 19, 542, 1970.
42. **Lenny, J. F. and Klemner, H. W.,** Factors controlling sexual reproduction and growth in *Pythium graninicola, Nature (London),* 209, 1365, 1966.
43. **Brack, A.,** Isolierung von Gentisalkohol neben Patulin, *Helv. Chim. Acta,* 30, 1, 1974.
44. **Foster, J. W., Woodruff, H. B., and McDaniel, L. E.,** Microbiological aspects of penicillin. III. Production of penicillin in surface cultures of *Penicillium notatum, J. Bacteriol.*, 46, 421, 1943.
45. **Emeh, C. O. and Marth, E. H.,** Cultural and nutritional factors that control rubratoxin formation, *J. Milk Food Technol.*, 39, 184, 1976.

46. **Vega, R. R. and LeTourneau, D.,** The effect of zinc on growth and sclerotial formation in *Whetzelinia sclerotium*, *Mycologia*, 66, 256, 1974.
47. **Sharp, E. L. and Smith, F. G.,** The influence of pH and zinc on vesicle formation in *Puccinia coronata* avenae corda, *Phytopathology*, 42, 581, 1952.
48. **Yamaguchi, H.,** Control of dimorphism in *Candida albicans* by zinc: effect on cell morphology and composition, *J. Gen. Microbiol.*, 86, 370, 1975.
49. **Hidy, P. H., Baldwin, R. S., Greasham, R. L., Keith, C. L., and McMullen, J. R.,** Zearalenone and some derivatives: production and biological activities, *Adv. Appl. Microbiol.*, 22, 59, 1977.
50. **Katz, E., Pienta, P., and Sivak, A.,** The role of nutrition in the synthesis of actinomycin, *Appl. Microbiol.*, 6, 236, 1958.
51. **Acker, R. F. and Lechavalier, H. A.,** Some nutritional requirements of *Streptomyces griseus* 3570 for growth and candicidin production, *Appl. Microbiol.*, 2, 152, 1954.
52. **Gallichio, V. and Gottlieb, D.,** The biosynthesis of chloramphenicol. III. Effects of micronutrients on synthesis, *Mycologia*, 50, 490, 1958.
53. **Basak, K. and Majumdar, S. K.,** Mineral nutrition of *Streptomyces kanamyceticus* for kanamycin formation, *Antimicrob. Agents. Chemother.*, 8, 391, 1975.
54. **Ristroph, J. D. and Ivins, B. E.,** Elaboration of *Bacillus anthracis* antigens in a new, defined culture medium, *Infect. Immun.*, 39, 483, 1983.
55. **Weinberg, E. D.,** Biosynthesis of bacitracin, gramicidin, and tyrocidine, in *Mechanism of Action and Biosynthesis of Antibiotics II*, Gottlieb, D. and Shaw, P. D., Eds., Springer-Verlag, Berlin, 1967, 240.
56. **Majumdar, S. K. and Bose, S. K.,** Trace element requirements of *Bacillus subtilis* for mycobacillin formation, *J. Bacteriol.*, 79, 564, 1960.
57. **Jansen, E. F. and Hirschmann, D. J.,** Subtilin — an antibacterial product of *Bacillus subtilis*. Culturing conditions and properties, *Arch. Biochem.*, 4, 297, 1944.
58. **Cooper, D. G., Macdonald, C. R., Duff, S. J. B., and Kosaric, N.,** Enhanced production of surfactin from *Bacillus subtilis* by continuous product removal and metal cation additions, *Appl. Environ. Microbiol.*, 42, 408, 1981.
59. **Weinberg, E. D.,** Manganese requirement for sporulation and other secondary biosynthetic processes of *Bacillus*, *Appl. Microbiol.*, 12, 436, 1964.
60. **Koo, F. C. W. and Peterson, J. W.,** The influence of nutritional factors on synthesis of *Salmonella* toxin, *J. Food Safety*, 3, 215, 1981.
61. **Maggon, K. K., Gupta, S. K., and Venkitasubramanian, T. A.,** Biosynthesis of aflatoxins, *Bacteriol. Rev.*, 41, 822, 1977.
62. **Steenbergen, S. T. and Weinberg, E. D.,** Trace metal requirements for malformin biosynthesis, *Growth*, 32, 125, 1968.
63. **Tinnell, W. J., Jefferson, B. L., and Benoit, R. E.,** Manganese-mediated morphogenesis in *Penicillium claviforme* and *Penicillium clavigerum*, *Can. J. Microbiol.*, 23, 209, 1977.
64. **Kubicek, C. P. and Rohr, M.,** Influence of manganese on enzyme synthesis and citric acid accumulation in *Aspergillus niger*, *Eur. J. Appl. Microbiol.*, 4, 167, 1977.
65. **Failla, M. L. and Weinberg, E. D.,** Zinc transport and metabolism by microorganisms, in *Zinc in the Environment II, Health Effects*, Nriagu, J. O., Ed., John Wiley & Sons, New York, 1980, 439.
66. **Hardie, L. P., Balkwill, D. L., and Stevens, S. E., Jr.,** Effects of iron starvation on the physiology of the cyanobacterium *Agmenellum quadruplicatum*, *Appl. Environ. Microbiol.*, 45, 999, 1983.
67. **Silver, S. and Jasper, P.,** Manganese transport in microorganisms, in *Microorganisms and Minerals*, Weinberg, E. D., Ed., Marcell Dekker, New York, 1977, 105.
68. **Failla, M. L.,** Zinc: functions and transport in microorganisms, in *Microorganisms and Minerals*, Weinberg, E. D., Ed., Marcell Dekker, New York, 1977, 151.
69. **Byers, B. R. and Arceneaux, J. E. L.,** Microbial transport and utilization of iron, in *Microorganisms and Minerals*, Weinberg, E. D., Ed., Marcell Dekker, New York, 1977, 215.
70. **Murphy, J. R., Michel, J. L., and Teng, M.,** Evidence that the regulation of diphtheria toxin production is directed at the level of transcription, *J. Bacteriol.*, 135, 511, 1978.
71. **Tejwani, G. A., Pedrosa, F. O., Pontremoli, S., and Horecker, B. L.,** Dual role of Zn^{2+} as inhibitor and activator of fructose 1,6-bisphosphatase of rat liver, *Proc. Natl. Acad. Sci., U.S.A.*, 73, 2692, 1976.
72. **Castric, K. F., McDevitt, D. A., and Castric, P. A.,** Influence of aeration on hydrogen cyanide biosynthesis by *Pseudomonas aeruginosa*, *Curr. Microbiol.*, 5, 223, 1981.
73. **Castric, P. A.,** The metabolism of hydrogen cyanide by bacteria, in *Cyanide in Biology*, Vennesland, B. et al., Eds., Academic Press, London, 1981, 233.
74. **Martin, J. F.,** Control of antibiotic synthesis by phosphate, *Adv. Biochem. Eng.*, 6, 105, 1977.
75. **Weinberg, E. D.,** Comparative aspects of secondary metabolism in cell cultures of green plants, animals, and microorganisms, in *Secondary Metabolism and Differentiation in Fungi*, Bennett, J. and Ciegler, A., Eds., Marcell Dekker, New York, 1983, 73.

76. **Weinberg, E. D.,** Iron witholding: a defense against infection and neoplasia, *Physiol. Rev.*, 64, 65, 1984.
77. **Schroth, M. N. and Hancock, J. G.,** Disease-suppressive soil and root-colonizing bacteria, *Science,* 216, 1376, 1982.
78. **Teintze, M. and Leong, J.,** Structure of pseudobactin A, a second siderophore from plant growth promoting Pseudomonas B10, *Biochemistry,* 20, 6457, 1981.

INDEX

A

B

C

D

E

F

G

H

I

P

Q

R

S

T